江西理工大学优秀学术著作出版基金资助

柔顺、并联机构空间构型综合理论及智能控制研究

朱大昌　陈德海　冯文结　著

北 京
冶 金 工 业 出 版 社
2014

内 容 提 要

本书针对目前柔顺、并联机构空间构型综合及其智能控制领域所涉及的并联机构空间构型综合及智能控制、全柔顺并联机构空间构型综合及智能控制等问题，详细阐述了并联机构空间构型综合，少自由度并联机构奇异位形，全柔顺并联机构空间构型综合及刚度模态矩阵分析，柔顺、并联机构智能控制系统的设计原则与方法。

本书可供从事柔顺、并联机构学及其智能控制研究工作的人员、大专院校机构学及智能控制专业的师生以及有意于在空间微纳尺度超精密定位与加工领域发展的企业的相关人员参考。

图书在版编目(CIP)数据

柔顺、并联机构空间构型综合理论及智能控制研究/朱大昌，陈德海，冯文结著．—北京：冶金工业出版社，2013.9（2014.5 重印）

ISBN 978-7-5024-6386-1

Ⅰ.①柔… Ⅱ.①朱… ②陈… ③冯… Ⅲ.①机构学 Ⅳ.①TH111

中国版本图书馆 CIP 数据核字(2013)第 225757 号

出 版 人　谭学余

地　　址　北京北河沿大街嵩祝院北巷 39 号，邮编 100009

电　　话　(010)64027926　电子信箱　yjcbs@cnmip.com.cn

责任编辑　杨　敏　美术编辑　杨　帆　版式设计　杨　帆

责任校对　禹　蕊　责任印制　牛晓波

ISBN 978-7-5024-6386-1

冶金工业出版社出版发行；各地新华书店经销；三河市双峰印刷装订有限公司印刷

2013 年 9 月第 1 版，2014 年 5 月第 2 次印刷

148mm×210mm；6.75 印张；198 千字；204 页

25.00 元

冶金工业出版社投稿电话：(010)64027932　投稿信箱：tougao@cnmip.com.cn

冶金工业出版社发行部　电话：(010) 64044283　传真：(010)64027893

冶金书店　地址：北京东四西大街 46 号(100010)　电话：(010)65289081(兼传真)

（本书如有印装质量问题，本社发行部负责退换）

前　言

本书是国家自然科学基金资助项目“基于全柔顺并联支撑机构的空间微纳尺度超精密定位系统研究”（项目编号：51165009）、国家自然科学基金资助项目“纳米级精度微运动测量与柔顺机构微动平台微位移检测”（项目编号：51105077）及江西省自然科学基金资助项目“基于全柔顺并联支撑机构的空间微纳尺度超精密定位系统研究”（项目编号：20114BAB206008）的部分研究成果（以上研究项目依托单位为江西理工大学），主要是介绍柔顺、并联机构空间构型综合理论及其智能控制。

本书主要内容包括：

（1）以螺旋理论及空间几何约束条件对并联机构空间构型进行综合。在各种约束存在条件下，分析并联机构动平台相应运动轨迹形式，为多刚体的位形分析提供可参考依据；对具有3支链5关节的三自由度纯移动并联机器人的奇异性进行分析与总结，并采用线性几何学图解形式直观给出其处于奇异位形；提出并联机构动平台关联运动这一概念，并针对4-RRCR型并联机构的支链配置形式进行运动性质仿真分析研究；采用螺旋理论对少自由度并联机构的速度反解及奇异性进行分析。

（2）提出并联机构滑模变结构控制、模糊控制及同步控制技术，解决并联机构轨迹精确跟踪这一技术问题。

(3) 提出全柔顺并联机构空间构型综合方法。综合采用螺旋理论及空间三维结构拓扑优化设计方法，进行全柔顺并联机构空间构型综合及基于 Ansys 软件的全柔顺并联机构运动学、动力学特性仿真研究；提出全柔顺并联机构模态刚度矩阵的计算方法。

(4) 提出全柔顺并联机构振动控制策略。采用滑模变结构控制策略，建立全柔顺并联机构空间运动学、动力学模型，并进行控制系统的仿真研究。

由于作者的时间和水平有限，书中难免有不足和不妥之处，敬请读者批评指正，对此作者不胜感谢！

作 者

2013 年 6 月

目 录

1 并联机构构型综合与奇异位形分析

1.1 应用螺旋理论对并联机器人形位的分析与综合

本文采用螺旋理论对并联机器人的约束与运动问题进行了分析和总结。在各种约束存在条件下，分析了刚体平台相应的运动轨迹形式，为今后对多刚体的形位分析提供了可行的理论依据。

1.1.1 引言

与常规的串联机器人相比，并联机器人具有刚度大、精度高以及承载能力强等优点，但在实际运用中，由于六自由度并联机器人的运动空间小、机构设计复杂等原因，使得六自由度并联机器人的运用受到限制。因此，研究者的目光转向了少自由度并联机器人研究方面，但是由于少自由度并联机器人的机构设计涉及复杂的约束系统分析，在这方面，方跃法等[1~3]采用螺旋理论对支链的形式进行了分析和综合，并在此基础上提出了关于少自由度并联机器人支链可能采取的组成形式。Herve 等[4]采用群论的方法设计出具有 3 个移动自由度的并联机器人机构。Kong 等[5]采用螺旋理论综合分析了具有平移性质的并联机器人机构构型。所有这些研究方法都是从几何学出发、凭借直觉判断对并联机器人可能的形位进行分析，具有一定的局限性。

本文依据螺旋理论对各个支链所产生的作用于动平台的约束进行了系统的分析和综合，并给出了设计并联机器人支链结构所遵循的一般规律，为并联机器人系统的机构设计提供了可行的途径。

1.1.2 螺旋理论概述

螺旋理论形成于 19 世纪，1900 年，Ball 完成经典著作《旋量理论》，直到 1948 年 Dimenberg 在分析空间机构时，才再次运用了这个理论，此后，螺旋理论逐步为机构学所重视，并得到迅速发展[6]。

数学中，一个旋量可以同时表示空间的一组对偶矢量，在机构学分析中，这一组对偶矢量可以表示为方向和位置，也可以表示为速度和角速度，如果从刚体力学的角度来定义旋量，则旋量又可用来表示力和力矩这一对矢量，因此，包含有 6 个标量（两个矢量）的旋量，对于研究空间机构的运动和动力学分析是至关重要的。

我们可以用一个单位螺旋来表示一条直线在空间中的方向和位置，这个单位螺旋 $\hat{\$}$ 包含了该直线的方位，形式如下：

$$\hat{\$}=\begin{bmatrix}\boldsymbol{s}\\ \boldsymbol{s}_0+\lambda\boldsymbol{s}\end{bmatrix} \tag{1}$$

式中，$\boldsymbol{s}$ 为单位矢量，其方向为螺旋轴线方向，与所表示的直线方向一致；$\boldsymbol{s}_0=\boldsymbol{r}\times\boldsymbol{s}$，$\boldsymbol{r}$ 表示从坐标原点到直线上任意一点的矢径，$\boldsymbol{s}_0$ 则被称为直线到原点的线距；λ 为螺旋的节距。

当 $\lambda=0$ 时，式（1）简化为以下形式：

$$\hat{\$}=\begin{bmatrix}\boldsymbol{s}\\ \boldsymbol{s}_0\end{bmatrix}=\begin{bmatrix}\boldsymbol{s}\\ \boldsymbol{r}\times\boldsymbol{s}\end{bmatrix} \tag{2}$$

式（2）如果表示为一个运动形式，则为一个转动副，该转动副的轴线为 s，与坐标原点的矢径为 $\boldsymbol{r}$，$\boldsymbol{r}=0$ 则表示该转动副的轴线通过坐标原点。式（2）如果表示为力形式，则表示一个纯力形式，该力的作用线方向为 s，与坐标原点的矢径为 $\boldsymbol{r}$，$\boldsymbol{r}=0$ 则表示该纯力通过坐标原点。

当 $\lambda=\infty$ 时，式（1）简化为以下形式：

$$\hat{\$}=\begin{bmatrix}0\\ \boldsymbol{s}\end{bmatrix} \tag{3}$$

式（3）如果表示为一个运动形式，则为一个移动副，该移动副的移动方向为 s。式（3）如果表示为力形式，则表示为一个纯力偶形式，该力偶的作用线方向为 s。

由于组成支链的运动副的基本形式为转动副和移动副，其他复杂的运动副可以通过这两种运动副的组合构成，例如：球面副可以通过三个共点不共面的转动副构成；圆柱副可以通过两个共轴线的移动副、转动副构成等。因此，在研究支链对动平台的约束问题上，我们假设支链的构成只有转动副和移动副，并不失一般性。

1.1.3 反螺旋系统性质

运动的反螺旋是约束力的概念，表示了物体在三维空间所受到的约束。两个螺旋分别表示为 $\$$ 和 $\$_r$，若这两个螺旋满足以下关系：

$$\$ \circ \$_r = 0 \tag{4}$$

则 $\$$ 和 $\$_r$ 互为反螺旋，式（4）中，"∘"表示互易积，该式等同于以下形式：

$$\$ = (s_1, s_2, s_3, s_4, s_5, s_6),\ \$_r = (s_{r1}, s_{r2}, s_{r3}, s_{r4}, s_{r5}, s_{r6})$$

$$\$ \circ \$_r = s_4 s_{r1} + s_5 s_{r2} + s_6 s_{r3} + s_1 s_{r4} + s_2 s_{r5} + s_3 s_{r6} = 0 \tag{5}$$

两个互为反螺旋的螺旋几何关系如图 1 所示。

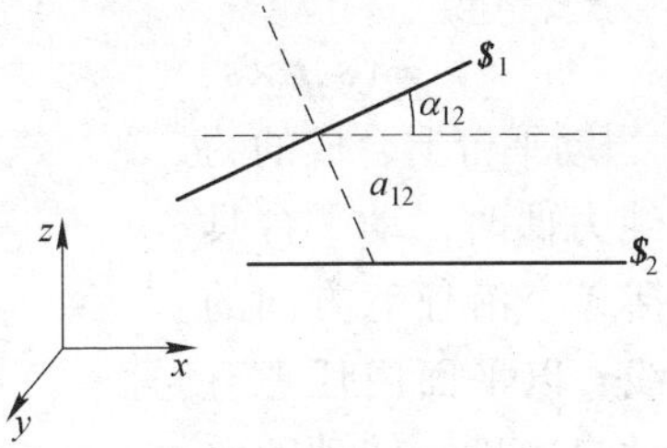

图 1　两个互为反螺旋之间的几何关系

设两螺旋的节距分别为 h_1 和 h_2，公法线长度为 a_{12}，方向矢量之间的夹角为 α_{12}，则两螺旋的互易积可表示为如下形式[7]：

$$(h_1 + h_2)\cos\alpha_{12} - a_{12}\sin\alpha_{12} = 0 \tag{6}$$

下面采用式（6）进行动平台的约束分析。

假设一个支链由 n（$n \leqslant 6$）个线性无关的运动副所组成，那么可以认为该支链为 n 阶螺旋系统，该支链对动平台产生的约束，即 n 阶螺旋系统的反螺旋系统为 $6-n$ 阶，也就是该支链作用在动平台上的约束数为 $6-n$ 个，表现为约束力形式，可以分为力、力偶以及两者的组合形式，简化为基本的两种形式（力和力偶）进行分析，并不失其一般性。

1.1.4 力偶约束分析

1.1.4.1 单个力偶约束分析

纯力偶的形式可以表示为：

$$\hat{\$}_r = (0,\ \boldsymbol{s}) \tag{7}$$

当物体受到这个力偶的作用时，表示物体没有作用线沿反螺旋轴线方向的角速度分量（s，0），否则功将不为0，所以任何轴线平行于 s 方向的转动都将被约束。另外，任何与 s 斜交的轴线都将对反螺旋产生转动分量，这种情况也将被约束。由于力偶在方向不变的情况下平行移动它的作用线并不改变其对动平台作用效果，所以，当作用在动平台上的力偶系方向一致的时候，该力偶系简化为单一的力偶作用。按照以上的分析可知，被约束的运动形式为：

$$\$ = (\boldsymbol{s}, \boldsymbol{r} \times \boldsymbol{s}) \tag{8}$$

式（8）说明，当动平台上合成的约束作用简化为一个纯力偶时，动平台具有5个自由度，包括3个沿任意方向的平移以及以 s 为法线、以平面内任意直线为轴线的两个转动自由度，转动的空间形式如图2所示。

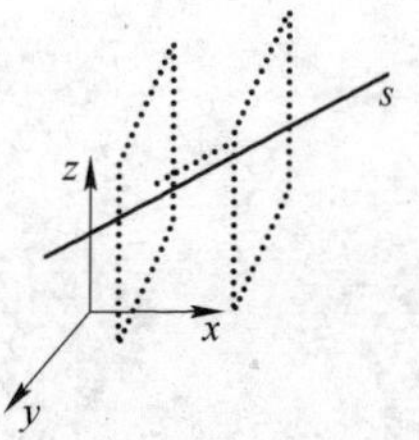

图2 单个力偶作用转动空间

1.1.4.2 两个力偶约束分析

两个纯力偶的形式可以表示为：

$$\hat{\$}_{r1} = (0, s_1),\ \hat{\$}_{r2} = (0, s_2) \tag{9}$$

动平台上作用的两个纯力偶形式如式（9）所示，由前述可知，两个力偶平移的结果必是相交的共面矢量，如果两个矢量共轴，则简化为第一种情况，若不共轴则两个纯力偶轴线相交确定一个平面，那么允许的运动形式为：

$$\$ = (s_1 \times s_2, \boldsymbol{r} \times (s_1 \times s_2)) = (s_1 \times s_2, (\boldsymbol{r} \times s_2) s_1 - (\boldsymbol{r} \times s_1) s_2) \tag{10}$$

式（10）说明，当动平台上合成的约束简化为两个共点共面的纯力偶时，动平台具有 4 个自由度，包括 3 个任意方向的平移以及与 2 个方向矢量决定平面的一族与法线平行的线系，转动的空间形式如图 3 所示。

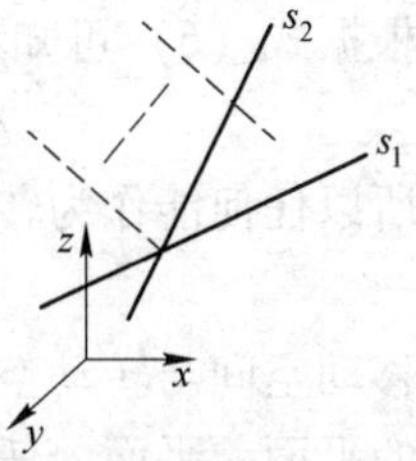

图 3 两个相交共面力偶转动空间

1.1.4.3 三个力偶约束分析

3 个纯力偶的形式可以表示为：

$$\hat{\$}_{r1} = (0,\ s_1),\ \hat{\$}_{r2} = (0,\ s_2),\ \hat{\$}_{r3} = (0,\ s_3) \qquad (11)$$

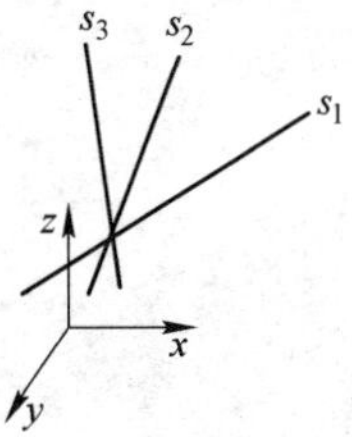

图 4 无转动空间

动平台上作用的 3 个纯力偶形式如式（11）所示，如 3 个力偶作用线共轴则简化为第一种情况；如 3 个力偶作用线共面则简化为第二种情况。当 3 个力偶作用线空间汇交时，其最大线性无关组为 3，限制了动平台三维空间的任意转动。由反螺旋约束条件可知，此时动平台具有在空间三个方向上任意平移的运动特性，其表现形式如图 4 所示。

1.1.5 力约束分析

1.1.5.1 单个力作用约束分析

当动平台存在纯力约束时，动平台的移动被约束的同时，其上的某些方向的转动同时被约束，这就产生了不完全的自由度。单个纯力约束与力偶约束不同之处是：纯力的作用点不能做空间平移。单个纯力约束的形式可以表示为：

$$\hat{\$}_r = (\boldsymbol{s}_r, \boldsymbol{r} \times \boldsymbol{s}_r) \qquad (12)$$

由于在与反螺旋线矢方向平行或相交的作用线上，允许的运动螺旋的节距都为 0，所以，当以这些作用线为转轴转动时在纯力线矢方向上没有速度分量。当动平台上作用的力线矢系共轴时，动平台具有 5 个自由度，包括 3 个方向的转动和 2 个方向的移动。移动方向为以 s 为法线的平面内任意移动，转动轴线由式（13）和式（14）所确定。

根据式（6）可知，当两线矢平行或相交时，存在以下关系：

$$\$ \circ \$_r = -a_{12}\sin\alpha_{12} = 0 \tag{13}$$

所以任何能作为转动轴线的线矢量必须满足以下关系式：

$$s_r(r' \times \boldsymbol{s}) + \boldsymbol{s}(\boldsymbol{r} \times \boldsymbol{s}_r) = 0 \tag{14}$$

移动空间与转动空间形式如图5所示。图中，P 代表一族以 $\boldsymbol{s}$ 为法线的平面，平面内任何移动为允许运动；r_1 代表以与方向矢量 $\boldsymbol{s}$ 上任何一点相交的直线为轴线的允许转动运动；r_2 代表以与方向矢量平行的直线矢量系为轴线的所允许的转动运动。

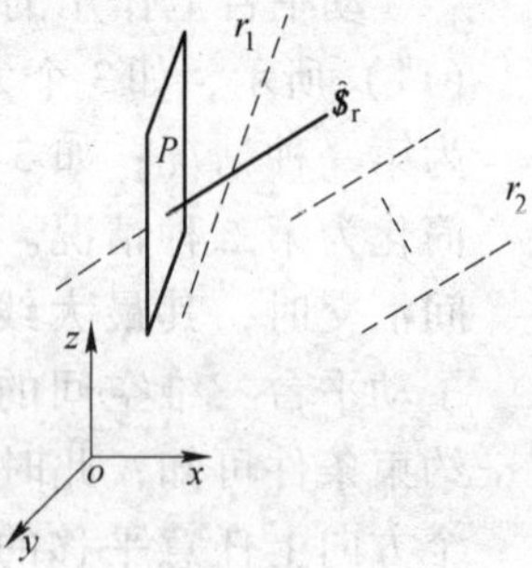

图5 单个力线矢的移动空间和转动空间

1.1.5.2 双线矢力作用约束分析

当动平台上有双线矢力约束作用时，该双线矢力的轴线为空间不共面、空间交错分布。由此两个空间交错的约束力线矢可约束动平台的两个移动方向，所允许的运动方向为同时与这两个线矢垂直的方向，这两个纯力线矢形式可以表示为：

$$\begin{aligned}\hat{\$}_{r1} &= (s_{r1}, r_1 \times s_{r1}) = (L_1, M_1, N_1, P_1, Q_1, R_1)\\ \hat{\$}_{r2} &= (s_{r2}, r_2 \times s_{r2}) = (L_2, M_2, N_2, P_2, Q_2, R_2)\end{aligned} \tag{15}$$

由于该两线矢力为空间交错形式，所以允许的移动形式可确定为：

$$\$ = (\boldsymbol{s}, \boldsymbol{r} \times \boldsymbol{s}) \tag{16}$$

式中，$s = \left(\dfrac{M_1N_2 - M_2N_1}{L_1M_2 - L_2M_1}, \dfrac{L_1N_2 - L_2N_1}{L_2M_1 - L_1M_2}, 1\right)$。

允许的转动方向可确定如下：

（1）与两线矢同时相交的可作为转动轴线；

（2）与 $\hat{\$}_{r1}$ 相交同时与 $\hat{\$}_{r2}$ 平行的可作为转动轴线；

（3）与 $\hat{\$}_{r2}$ 相交同时与 $\hat{\$}_{r1}$ 平行的可作为转动轴线。

与两线矢同时相交的轴线组成的区域如图 6 所示，所有以 $\hat{\$}_{r1}$ 和 $\hat{\$}_{r2}$ 线矢为顶点的线矢都可以作为转动轴线。

以第二种情况分析，与 $\hat{\$}_{r2}$ 平行的平面可表示为两个线性无关的基平面，所有与 $\hat{\$}_{r2}$ 平行的平面都可由这两个基平面的组合表示。两个线性无关的基平面表示为：

$$r_{p1}s_{p1} = s_{0p1},\ r_{p2}s_{p2} = s_{0p2} \tag{17}$$

式中，$s_{p1} = \left(-\dfrac{N_2}{L_2},0,1\right)$，$s_{p2} = \left(-\dfrac{M_2}{L_2},1,0\right)$。

组合平面的形式可以表示为：$k_1(r_{p1}s_{p1}) + k_2(r_{p2}s_{p2}) = k_1s_{0p1} + k_2s_{0p2}$

标准形式为：

$$r_ps_p = s_{0p} \tag{18}$$

线矢 $\hat{\$}_{r1}$ 与平行平面的交点坐标形式表示为：

$$r(ss_p) = s_p \times s_{01} + s_{0p}s \tag{19}$$

因此，当直线线矢同时满足式（18）与式（19）时，可以作为转动轴线，如图 7 所示。

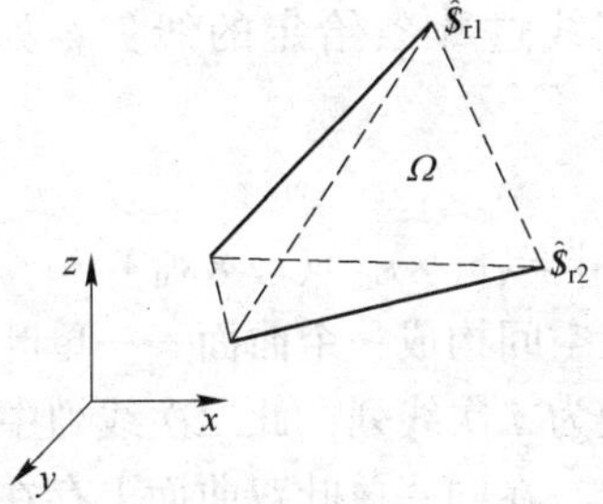

图 6 同时与两线矢力相交的转动轴线的确定（Ω）

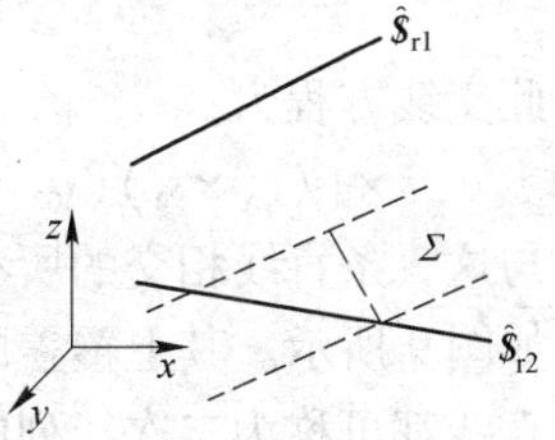

图 7 与两线矢力平行及相交的转动轴线确定（Σ）

1.1.5.3 三线矢力作用约束分析

三线矢力同时作用在动平台上，由于线矢力无保持其作用线平行移动不改变其作用效果的特性，所以在进行其约束分析时分为以下几种情况：

（1）三线矢力呈空间分布，相互线性无关，这种情况下，三线矢力限制了动平台在三维空间中的移动，并对转动有不完全约束作用；

（2）三线矢力空间共点，这种情况下，纯力线矢对动平台的三维空间移动进行约束，并对转动有不完全约束；

（3）三线矢力共面共点，三线矢力的最大线性无关组减为2，对所在平面的任意移动进行约束，虽对转动有不完全约束，但是动平台的自由度数增加为4，产生1个平移冗余自由度；

（4）三线矢力共面不共点，此种情况下动平台具有3个自由度，1平移2转动；

（5）三线矢力空间平行，此时约束了两个方向的转动和一个方向的平移。

A 空间分布且线性无关[7]

3个线矢力限制了物体在三维空间的3个方向上的移动，同时对存在的转动也有一定的约束，物体允许的转动轴线必须能与3个线矢力的作用线同时相交。设3个线性无关的线矢力为：$\hat{\$}_1$、$\hat{\$}_2$、$\hat{\$}_3$，在$\hat{\$}_1$上任意取一点$A$，该点和另外两个线矢组成了两个相交平面，由点$A$的选取原则可知，两平面的交线过3条给定的线矢，如图8所示。

此交线方程为：

$$r \times [(r_A \times s_1)(s_{02} - r_A \times s_2)] = (r_A \times s_{02})(r_A \times s_{01}) \qquad (20)$$

与这3条直线相交的所有直线在空间构成一个曲面——单叶双曲面，如图9所示。其上覆盖的直线族为二次线列，此二次线列中的每一条直线都可称为二次线列的发生线，在同一单叶双曲面上存在两个二次线列，由两族直线构成，这两族直线构成同一表面并完全覆盖此表面，一族直线的每条线必与另一族的每一条直线相交，而不与本族的任何直线相交。

B 三线矢力共面共点

共点共面的3个线矢力，其最大的线性无关组为2，即其中任意一个线矢力可由另外两个线矢力线性表示，约束了空间中的三维任意

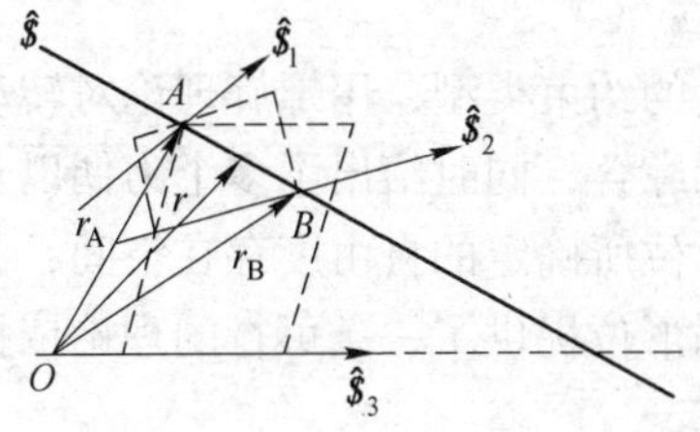

图 8 与三条线矢力同时相交的螺旋

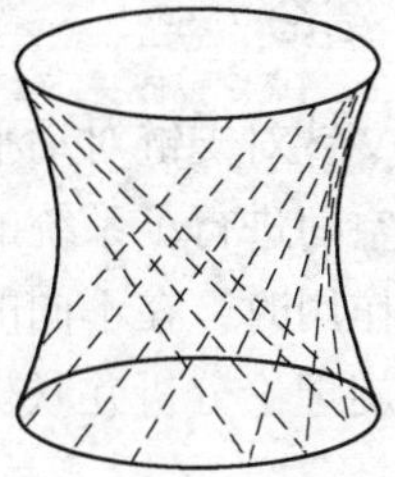

图 9 两族二次线列

移动，同时要求任何转动在 3 个线矢力方向上无移动分量。根据以上分析可知，平面内的任意直线可作为转动轴线，并且空间中任何通过三线矢力交点的直线也可作为转动轴线，转动区域标识为 Ω，移动区域标识为 Σ，则三线矢力共面共点的转动空间与移动空间如图 10 所示。

三线矢力共面不共点的情况与共面共点相类似，移动空间不变，转动空间只为 Ω_1。

C 三线矢力空间平行约束分析

空间平行的 3 个线矢力约束了沿线矢力方向的移动自由度。根据以上分析可知，此三线矢力允许的转动轴线为与三线矢力平行方向，任何与该线矢力平行的直线都可作为转动轴线，其转动空间与移动空间如图 11 所示。

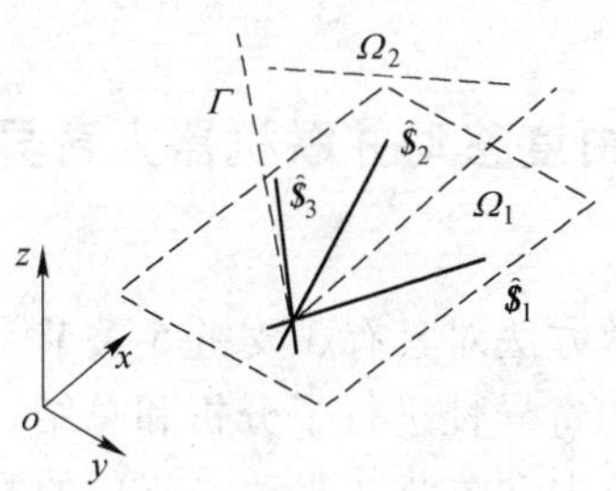

图 10 三线矢力共面共点的转动及移动空间

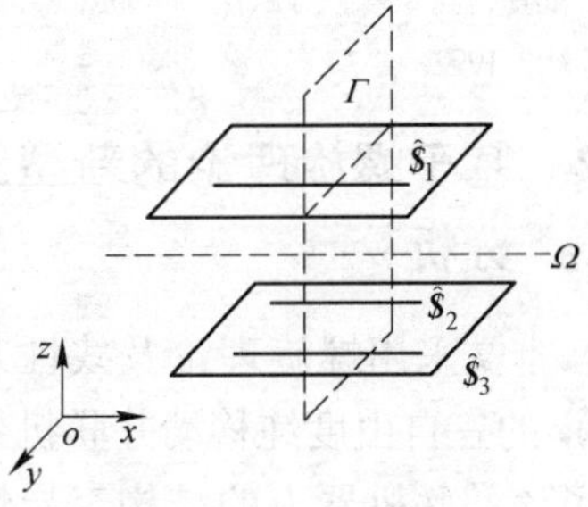

图 11 三线矢力平行的转动及移动空间的确定

1.1.6 结论

本文针对支链对动平台产生的约束类型，用螺旋理论对转动空间和移动空间进行了系统的分析和综合，同时给出了多个力偶以及多个线矢力作用时，在不同的相对形位所确定的自由度存在空间，为今后应用螺旋理论分析并联机器人的形位提供了一条可行的理论依据。

参 考 文 献

[1] Fang Y F, Tsai L W. Analytical identification of limb structures for translational parallel manipulators [J], Journal of Robotic Systems, 2004, 21 (5): 209 ~218.

[2] Fang Y F, Tsai L W. Structure synthesis of a class of 3-DoF rotational parallel manipulators [J]. IEEE Transaction on Robotics and Automation, 2004, 20 (1): 117 ~121.

[3] Fang Y F, Tsai L W. Structure synthesis of a class of 4-DoF and 5-DoF parallel manipulators with identical limb structures [J]. The International Journal of Robotics Research, 2002, 21 (9): 799 ~810.

[4] Herve J M, Sparacino E. Structural synthesis of parallel robots generating spatial translation [C] //Proceedings of the IEEE International Conference on Robotics and Automation. USA: IEEE, 1992, 808 ~813.

[5] Kong X, Gosselin C M. Generation of parallel manipulators with three translational degree of freedom based on screw theory [C] //Proceedings of 2001 CCToMM Symposium on Mechanisms, Machines and Mechantronics. http: //www. cim. mcgill. calexvit/sM3/Papers/M3-01-012. pdf, 2001.

[6] Hunt K H. Kinematic Geometry of Mechanisms [M]. London, England: Oxford University Press, 1978.

[7] 黄真，孔令富，方跃法. 并联机器人机构学理论及控制 [M]. 北京：机械工业出版社，1997.

1.2 基于螺旋理论的新型三自由度移动并联机器人奇异性分析

本文采用螺旋理论及线性几何学方法对具有 3 支链 5 关节（P-4R）的三自由度纯移动并联机器人的奇异性进行了分析和总结。研究了该并联机器人的结构奇异性以及主动关节（驱动关节）的配置奇异性，并采用线性几何学图解形式直观地给出其处于奇异时的形位，为进一步研究其动力学、运动学性质奠定了基础。

1.2.1 引言

自从 Gough 第一次将并联机器人引入到疲劳实验中并被 Stewart 用于飞行模拟器[1]之后，越来越多的研究者对并联机构进行了研究和探讨。由于并联机构具有小惯性、高承载力、高刚度以及良好的动态特性，其结构可用于非常多的领域（从机器人到并联机床），然而在很多场合并不需要用到六自由度并联机构，而少自由度并联机构尤其是三自由度并联机构应用越来越为广泛。Fang[2]采用反螺旋理论对一系列三自由度转动并联机器人可能的支链结构进行了列举，归纳总结了可行的支链列表，这些支链可以用于三自由度转动并联机器人的结构组成。Joshi[3]列举了四支链三自由度的并联机构，并针对该类并联机构提出了一种新的方法，对其运动学进行分析。Kong[4]基于螺旋理论，提出了一种对三自由度移动机构进行分析和总结的方法，并对该类机构主动关节（驱动关节）的有效性条件进行了分析。Miller[5]研究了三自由度电动机在工作空间轨迹中的影响。还有许多文献涉及三自由度并联机构的研究与分析[6~8]。

文献［8］提出了一种关于三自由度移动机器人的新机构，动平台通过 3 根支链与基座相连接，每个支链由 4 个两两平行的转动副和 1 个作为驱动装置的移动副组成，该移动副可配置在支链上的任意位置。

本文分析了具有 3 个转动自由度的 P-4R 并联机器人的主动奇异（驱动奇异）和形位奇异（结构奇异）问题，并以线性几何学图解形式给出了机构奇异特征，为该类型并联机器人的奇异性分析做了较为详尽的理论探讨。

1.2.2 反螺旋理论概述

一个单位螺旋可以定义为 6×1 矩阵形式[9]：

$$\boldsymbol{\$}_i = \begin{pmatrix} \boldsymbol{s}_i \\ \boldsymbol{r}_i \times \boldsymbol{s}_i + \lambda \boldsymbol{s}_i \end{pmatrix} \tag{1}$$

式中，$\boldsymbol{s}_i$ 为沿螺旋轴线方向的单位矢量；$\boldsymbol{r}_i$ 为从坐标原点指向螺旋轴线上的任意一点的线矢量；λ 为螺旋的节距；$\boldsymbol{r}_i \times \boldsymbol{s}_i$ 定义了沿螺旋方向的运动。

对于转动关节，$\lambda=0$，螺旋形式表示为：

$$\hat{\$}_i=(\boldsymbol{s}_i,\boldsymbol{r}_i\times\boldsymbol{s}_i)^{\mathrm{T}} \tag{2}$$

对于移动关节，$\lambda=\infty$，螺旋形式表示为：

$$\hat{\$}_i=(0,s_i)^{\mathrm{T}} \tag{3}$$

当两个螺旋互为反螺旋时，应满足下列条件：

$$\hat{\$}_r^{\mathrm{T}}\circ\hat{\$}_i=0 \tag{4}$$

式中，$i=1, 2, \cdots, 6-n$，$r=1, 2, \cdots, n$。

螺旋的转置可以定义为：$\hat{\$}_r^{\mathrm{T}}=[s_{r4}\quad s_{r5}\quad s_{r6}\quad s_{r1}\quad s_{r2}\quad s_{r3}]$

式（4）可以表示为：

$$\hat{\$}_r^{\mathrm{T}}\circ\hat{\$}_i=s_{r4}s_1+s_{r5}s_2+s_{r6}s_3+s_{r1}s_4+s_{r2}s_5+s_{r3}s_6 \tag{5}$$

式中，$s_{ri}(i=1,2,\cdots,6)$ 为第 r 个单位螺旋中的第 i 个分量，$\hat{\$}_i(i=1,2,\cdots,6)$ 为 $\hat{\$}_r$的反螺旋形式。

1.2.3 3P-4R 型移动并联机器人奇异分析

并联机器人的奇异分析在机构设计、轨迹规划及控制方面有着极其重要的作用。如果机构的雅克比矩阵不满秩，则称该机构处于奇异。这里主要分析该类3P-4R 移动并联机器人的两种奇异性：形位奇异以及主动奇异。

1.2.3.1 形位奇异分析

该机构动平台由 3 根对称相同的支链与下平台相连接。

如图 1 所示，每个支链由 4 个从动转动副和一个主动移动副组成，该移动副可以装在支链的任意部位，为了分析方便，只考虑移动副配置在下平台上。图 2 所示为第 i 个支链的运动学模型，4 个转动关节分别为 R_{i1}，R_{i2}，R_{i3}，R_{i4}，移动关节为 P_i。R_{i1}与 R_{i2}的轴线以及 R_{i3}与 R_{i4}的轴线相互平行。

假设 R_{i1}与 R_{i2}的轴线方向表示为：$\boldsymbol{s}_{i1}=(m_{i1},n_{i1},p_{i1})$，同理，设 R_{i3}与 R_{i4}的轴线方向为：$\boldsymbol{s}_{i2}=(m_{i2},n_{i2},p_{i2})$。移动副 P_i 的轴线方向为：$\boldsymbol{s}_{ip}=(x_i,y_i,z_i)$。从坐标原点指向各个运动副中心的线矢量表示为：$\boldsymbol{r}_{im}=(x_{im},y_{im},z_{im})(m=1,2,3)$，对于第 i 根支链，建立螺旋系统，有如下形式：

$$\$_{pi} = (0, \boldsymbol{s}_{pi}),\ \$_{i,r1} = (\boldsymbol{s}_{i1}, \boldsymbol{r}_{i,r1} \times \boldsymbol{s}_{i1}),\ \$_{i,r2} = (\boldsymbol{s}_{i1}, \boldsymbol{r}_{i,r2} \times \boldsymbol{s}_{i1}),$$
$$\$_{i,r3} = (\boldsymbol{s}_{i2}, \boldsymbol{r}_{i,r3} \times \boldsymbol{s}_{i2}),\ \$_{i,r4} = (\boldsymbol{s}_{i2}, \boldsymbol{r}_{i,r4} \times \boldsymbol{s}_{i2}) \tag{6}$$

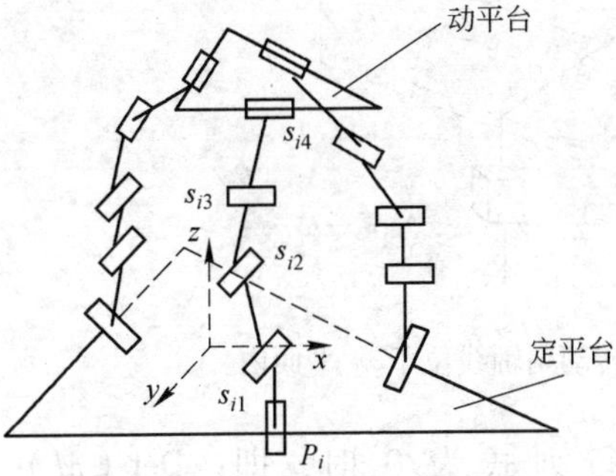

图 1 P-4R 型并联机构构型

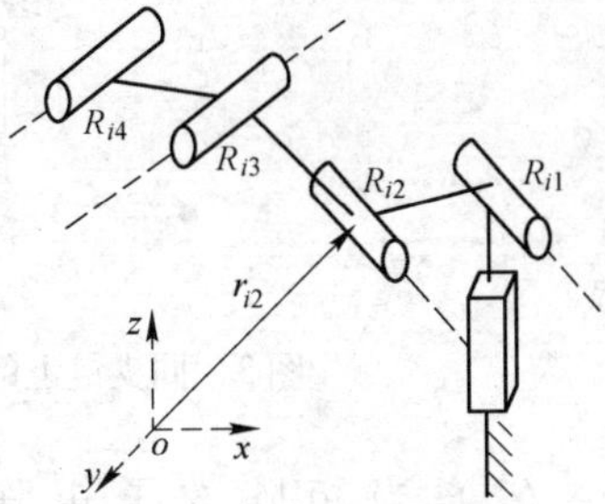

图 2 P-4R 型并联机构支链运动学模型

等式（4）的反螺旋形式为：

$$\hat{\$}_i = \left(0,0,0,\frac{n_{i2}p_{i1} - n_{i1}p_{i2}}{m_{i2}n_{i1} - m_{i1}n_{i2}},\frac{m_{i1}p_{i2} - m_{i2}p_{i1}}{m_{i2}n_{i1} - m_{i1}n_{i2}},1\right) \tag{7}$$

考虑 3 个支链的反螺旋所组成的反螺旋系统，可以得到一个 3 × 3 的矩阵形式：

$$\text{Matrix} = [N_i, M_i, 1] \quad i = 1, 2, 3 \tag{8}$$

式中，$N_i = \dfrac{n_{i2}p_{i1} - n_{i1}p_{i2}}{m_{i2}n_{i1} - m_{i1}n_{i2}}$，$M_i = \dfrac{m_{i1}p_{i2} - m_{i2}p_{i1}}{m_{i2}n_{i1} - m_{i1}n_{i2}}$。

结构的奇异性有以下 3 种情况：

（1）当 $m_{i2}n_{i1} - m_{i1}n_{i2} = 0$ 时，所有的转动副轴线位于两个互相平行的平面内，等式（7）的形式变为：

$$\hat{\$}_i = \left(0,0,0,-\frac{n_{i2}}{m_{i2}},1,0\right) \tag{9}$$

很明显，3 个反螺旋中必定有一个反螺旋与另外两个反螺旋线性相关，式（9）的最大无关数为 2，此种情况下，机构的自由度数目增加至 4，3 个平动 1 个转动，3 支链对动平台的约束退化为约束平台在 xy 平面内的转动，如图 3 所示。

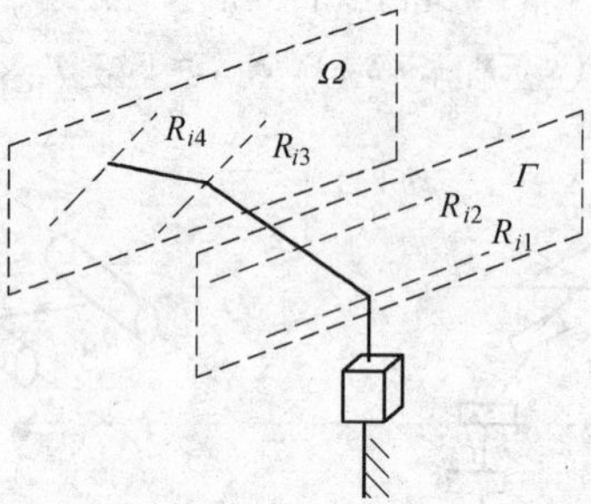

图 3 同支链上的转动副轴线位于 xy 平面内

（2）当矩阵的子矩阵行列式为 0 时，即：Det（M）= $\begin{vmatrix} \dfrac{n_{12}p_{11}-n_{11}p_{12}}{m_{12}n_{11}-m_{11}n_{12}} & \dfrac{m_{11}p_{12}-m_{12}p_{11}}{m_{12}n_{11}-m_{11}n_{12}} \\ \dfrac{n_{22}p_{21}-n_{21}p_{22}}{m_{22}n_{21}-m_{21}n_{22}} & \dfrac{m_{21}p_{22}-m_{22}p_{21}}{m_{22}n_{21}-m_{21}n_{22}} \end{vmatrix}=0$，说明反螺旋两两线性相关，此种情况下产生结构的奇异性如图 4 所示。

（3）当矩阵秩为 0 时，即：$\text{Det}(Matrix)=|N_i,M_i,1|=0, i=1,2,3$，表示为：

$$M_2N_1+M_3N_2+M_1N_3-M_3N_1-M_1N_2-M_2N_3=0 \tag{10}$$

转动副轴线的配置只要满足式（10）都将产生奇异性，如图 5 所示。

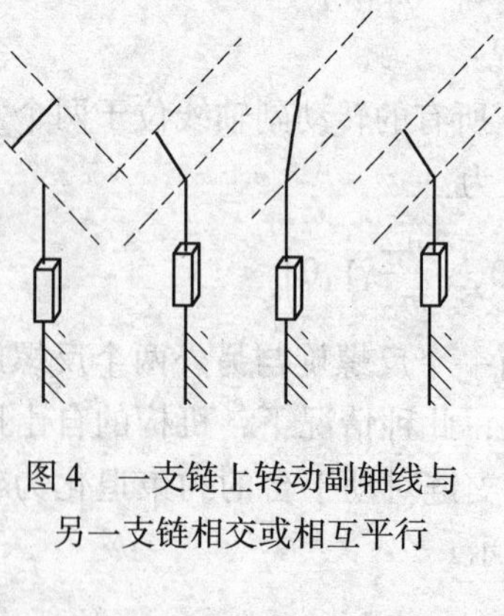

图 4 一支链上转动副轴线与另一支链相交或相互平行

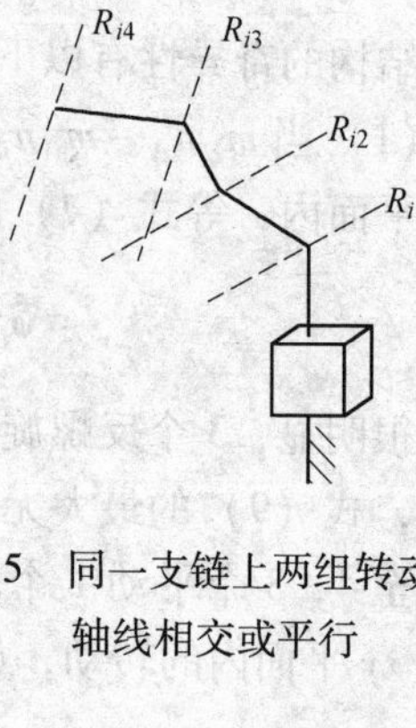

图 5 同一支链上两组转动副轴线相交或平行

1.2.3.2 主动奇异性

将移动副作为主动关节的三自由度并联机器人是 Gosselin[10] 提出的，将其作为机器人的手腕及操纵杆。主动奇异包括了主动配置奇异以及运动学奇异两种类型，这里只限于讨论将三自由度 P－4R 型并联机器人的移动副作为主动关节的配置奇异问题，该移动副配置在下平台上。

设移动副主动关节轴线与 xy 平面的夹角为 θ_i，与 x 轴的夹角为 α_i，则由所有的主动关节轴线所组成的矩阵可以表示为：

$$\boldsymbol{A}=\begin{pmatrix} c\theta_1 c\alpha_1 & c\theta_1 s\alpha_1 & s\theta_1 \\ c\theta_2 c\alpha_2 & c\theta_2 s\alpha_2 & s\theta_2 \\ c\theta_3 c\alpha_3 & c\theta_3 s\alpha_3 & s\theta_3 \end{pmatrix} \tag{11}$$

式中，$c[\cdot]=\cos[\cdot]$，$s[\cdot]=\sin[\cdot]$。

当矩阵的秩为 0 时,则发生主动奇异：

$$\begin{aligned}\mathrm{Det}(\boldsymbol{A}) = {} & c\alpha_1 c\theta_1 c\theta_2 s\alpha_2 s\theta_3 + c\alpha_2 c\theta_2 c\theta_3 s\alpha_3 s\theta_1 + c\alpha_3 c\theta_1 c\theta_3 s\alpha_1 s\theta_2 - {} \\ & c\alpha_3 c\theta_2 c\theta_3 s\alpha_2 s\theta_1 - c\alpha_1 c\theta_1 c\theta_3 s\alpha_3 s\theta_2 - c\alpha_2 c\theta_1 c\theta_2 s\alpha_1 s\theta_3 \end{aligned} \tag{12}$$

（1）当每个主动关节的 α_i 相等时，式（12）为 0，此时发生主动奇异，如图 6 所示。

（2）如果 $\theta_i=0$，也就是说所有的主动关节轴线位于 xy 平面内，矩阵的第三列全部为 0，此时矩阵的行列式为 0，这种情况下也将产生主动奇异，如图 7 所示。

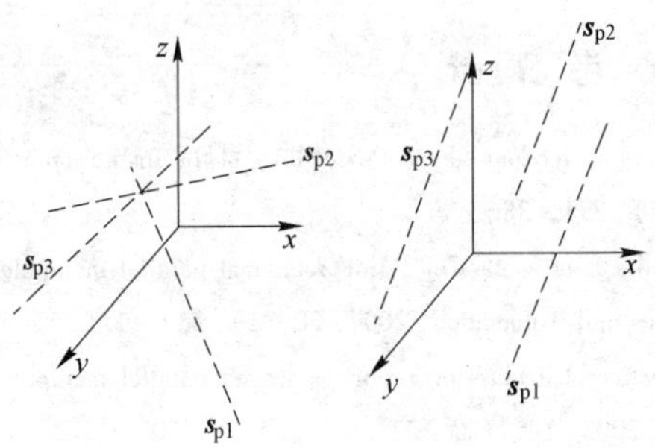

图 6 主动关节轴线在空间相交或平行

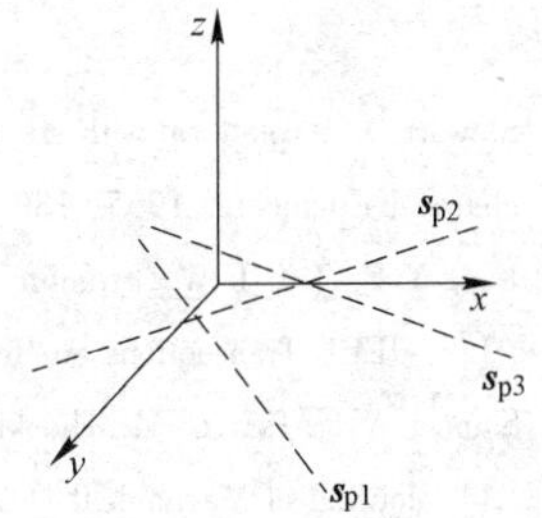

图 7 三个主动关节轴线位于同一平面内

1.2.4 3P-4R 并联机器人奇异性图解

如前面所分析，在该并联机器人中存在两种奇异性：结构奇异和主动奇异。

结构奇异取决于支链上的各个关节轴线在运动中所处的特殊形位，这些特殊形位包括：支链上 4 个转动副轴线两两位于两个相互平行的平面内；两对转动副轴线两两空间相交；各支链之间的转动副轴线相交或平行。主动奇异发生在作为主动关节的移动副轴线配置的特殊位置，这些特殊位置包括：2 个或 3 个主动关节轴线在空间相交或平行；3 个主动关节轴线位于同一个平面内（xy）。所有奇异形位如图 5 ~ 图 7 所示。

1.2.5 结论

通过采用螺旋和反螺旋理论及线性几何方法，对一种新型的三自由度 P-4R 型平移并联机器人的结构奇异形位及主动奇异形位进行了系统的分析和综合。由于该并联机器人为三自由度平移机构，所以在奇异性分析时，只需要分析其矩阵相关问题。本文分析确定了 3 种结构奇异形位及 2 种主动奇异形位，并给出了各类奇异形位的图解形式，这些奇异形位的分析有助于在设计该类并联机器人时避免其奇异形位的产生，同时为螺旋和反螺旋及线性几何方法在并联机器人奇异性分析方面的运用做了进一步的研究和探讨。

参 考 文 献

[1] Stewart D. A platform with six degrees of freedom [J]. Proceedings of the Institution of Mechanical Engineers, 1965, 180 (15): 371 ~ 386.

[2] Fang Y F, Tsai L W. Structure synthesis of a class of 3-DoF rotational parallel manipulators [J]. IEEE Transactions on Robotics and Automation, 2004, 20 (1): 117 ~ 121.

[3] Sameer A J, Tsai L W. The kinematics of a class of 3-DoF, 4-legged parallel manipulators [J]. Journal of Mechanical Design, 2003, 125 (1): 52 ~ 60.

[4] Kong X W, Gosselin C M. Type synthesis of 3-DoF translational parallel manipulators based on screw theory [J]. Journal of Mechanical Design, 2004, 126 (1): 83 ~ 92.

[5] Miller K. Maximization of workspace volume of 3-DoF spatial parallel manipulators [J].

Journal of Mechanical Design, 2002, 124 (2): 347 ~358.
[6] Li Y W, Wang J S, Wang L P, et al. Inverse dynamics and simulation of a 3-DoF spatial parallel manipulator [C] //Proceedings of the IEEE International Conference on Robotics and Automation. USA: IEEE Press, 2003: 4092 ~4097.
[7] Wang J, Gosselin C M. Singularity loci of a special class of spherical 3-DoF parallel mechanisms with prismatic actuators [J]. Journal of Mechanical Design, 2004, 124 (2): 319 ~327.
[8] Marco C, Vincenzo P C. Position analysis of a new family of 3-DoF translational parallel manipulators [J]. Journal of Mechanical Design, 2003, 125 (2): 316 ~322.
[9] Tsai L W. Robot analysis, the mechanics of serial and parallel manipulators [M]. New York: Wiley, 1999.
[10] Gosselin C M, Lavoie E. On the kinematic design of spherical 3-DoF parallel manipulators [J]. The International Journal of Robotics Research, 1993, 12 (4): 394 ~402.

1.3 4-RRCR 型并联机构动平台关联运动特性分析

4-RRCR 型并联机构支链采用了两组转动副形式，与定平台相连接的一组转动副的轴线相互平行，另一组转动副的轴线则相互交于空间的一个公共点。在这种配置情况下，动平台获得了两个关联移动关联运动特性，所设计的并联机构的运动形式发生了改变。文中对这种运动形式进行分析，提出了关联运动概念用以解决这种由于设计所造成的运动性质发生改变的情况，并通过仿真对比加以说明。

1.3.1 引言

与常规的串联机器人相比，并联机器人具有刚度大、精度高以及承载能力高等优点，但在实际运用中，由于六自由度并联机器人的运动空间小、复杂的机构设计等原因，使得六自由度并联机器人的运用受到限制。因此，研究者的目光转向了少自由度并联机器人研究领域。方跃法等[1~3]采用螺旋理论对支链的形式进行了分析和综合，并在此基础上提出了关于少自由度并联机器人支链所可能采取的组成形式。文献［3，4］采用了螺旋理论对 4－RRCR 型并联机构支链约束系统进行了分析并提出：为了使得该机构具有 3R1T 的运动特性，其支链中与定平台相连接的两个 R（Rotational Joint）副的轴线相互平行，与动平台相连接的 C（Cylindrical Joint）副与 R 副的轴线则全部相交于空间的一个公共点。这种配置使得该并联机构所产生的两组纯力约束作用在相交点[5,6]，并在力矢量所决定的平面内对动平台的两个移动方向进行约束，从而达到动平台具有 3R1T 的运动特性。然而，文献［3］所设计的并联机构动平台由于相交点位置不同，出现了两个关联移动运动特性，而这种关联运动是在并联机构的应用及控制系统设计中所应避免的。在关联运动特性分析方面，程志红等[7]在平面连杆机构运动学分析中提出了机构运动中的关联矩阵方法，并定义了关联矩阵的归并计算。

本文基于文献［3］中采用螺旋理论对 4-RRCR 型并联机构运动性质的分析，针对相交点位置不同所引起的关联运动特性进行了分析并提出采用机构优化设计的方法避免并联机构关联运动的产生。通过对两种不同配置类型的 4-RRCR 型并联机构的 SimMechanics 的仿真验证了所提观点的正确性。

1.3.2 两种不同配置的 4-RRCR 型并联机构模型

根据文献［3，6］中的分析，只要是遵循了在设计过程中所规定的，也即是支链中的第一组与定平台相连接的转动副的轴线相互平行，另外的圆柱副以及与动平台相连接的转动副轴线都相交于空间的任意一点，那么该并联机构的动平台即为 3R1T 四自由度的运动形式，并指出支链所产生的约束形式为纯力约束矢量，该约束力矢量通过相交点并与对应的支链中轴线相互平行的转动副轴线方向相一致。本处提出了两种 4-RRCR 型并联机构，此两种并联机构的共同点为经过螺旋理论的分析都具有 3R1T 的运动特性，但约束力矢量所位于的平面不同（随着相交点 P 位置的不同，该平面为一系列相互平行的平面族），如图 1 所示。

另一个机构模型为，副轴线方向相交于动平台的中心，如图 2 所示。

运用 Matlab 中的 SimMechanics 对上述两种机构建模并进行机构仿真。模型 1 的仿真结构如图 3 及图 4 所示。图 3 的纵坐标为动平台中心点的三个平移距离（单位：mm），图 4 的纵坐标为动平台中心的三个转动速度（单位：rad/s）。两图的横坐标为时间轴，其采样时

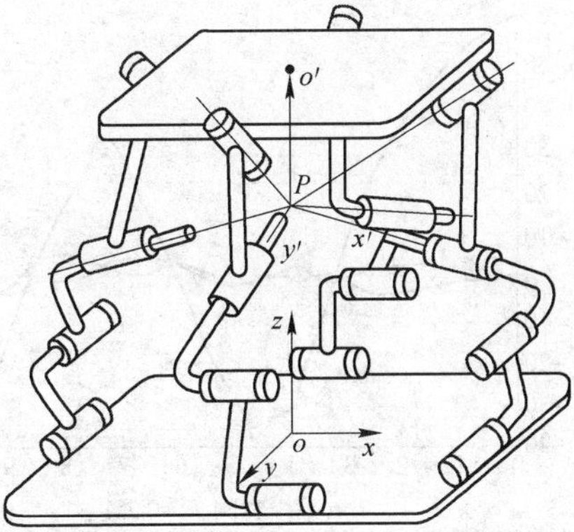

图 1 R、C 副轴线相交点在 z 轴上任意一点（模型 1）

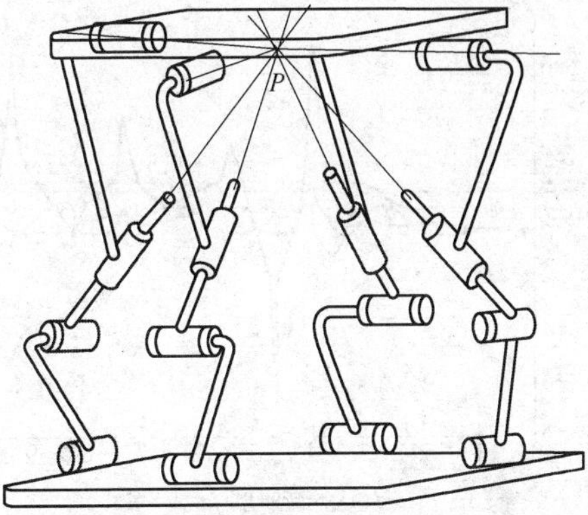

图 2 R、C 副轴线相交点在动平台中心（模型 2）

间 $T_s=0.2s$。

模型 2 的仿真结果如图 5 及图 6 所示。图 5 的纵坐标为动平台中心点的三个平移距离(单位:mm),图 6 的纵坐标为动平台中心的三个转动速度(单位:rad/s)。两图的横坐标为时间轴,其采样时间 $T_s=0.2s$。

在仿真过程中，驱动由与定平台相连接的 4 个转动关节所提供，模型 1 与模型 2 的驱动参数相同，分别为 $V_{R1}=V_{R2}=0.4rad/s$，$V_{R3}=0.3rad/s$，$V'_{R3}=0.5rad/s$，其中：V_{Ri} 为转动关节所提供的驱动转速，

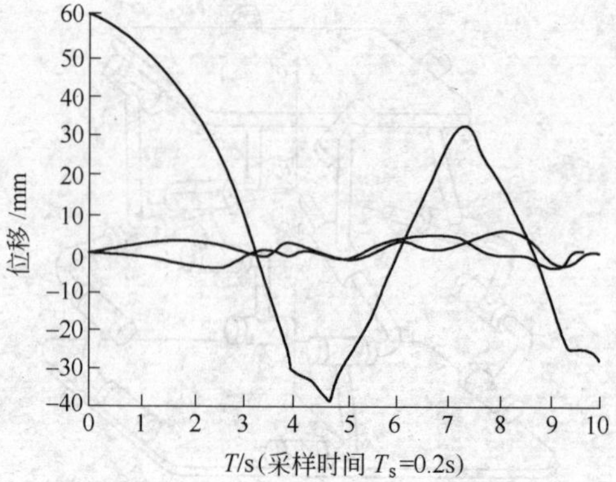

图 3　模型 1 中动平台的 3 个平移运动

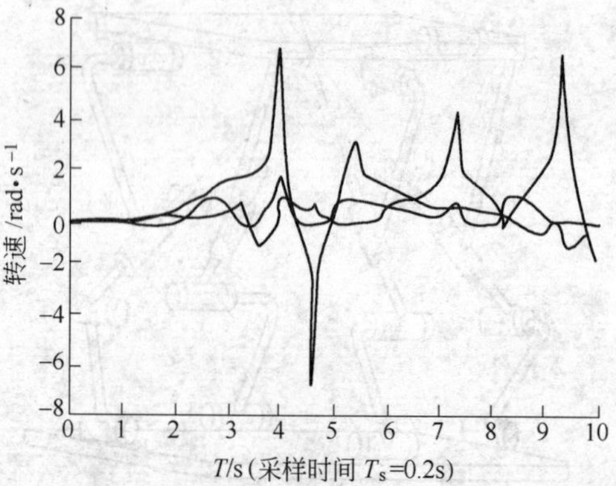

图 4　模型 1 中动平台的 3 个旋转运动

由 Matlab SimMechanics 中的 Joints initial conditions 模块所提供。

通过机构仿真对比可知，由于约束力矢量交点的不同，从而导致了动平台具有了关联运动的特性。

1.3.3　模型 1 动平台运动性质分析

关联运动与不完全自由度有着本质的区别：不完全自由度是由于受到约束而在某一特定的方向上不能任意运动，而本文提出的关联运

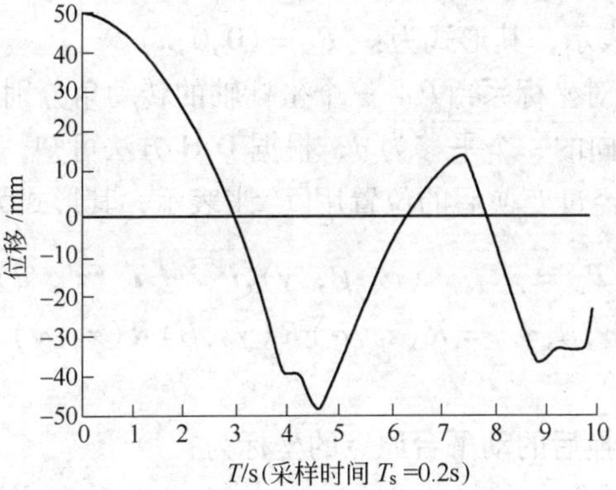

图 5　模型 2 中动平台的一个平移运动

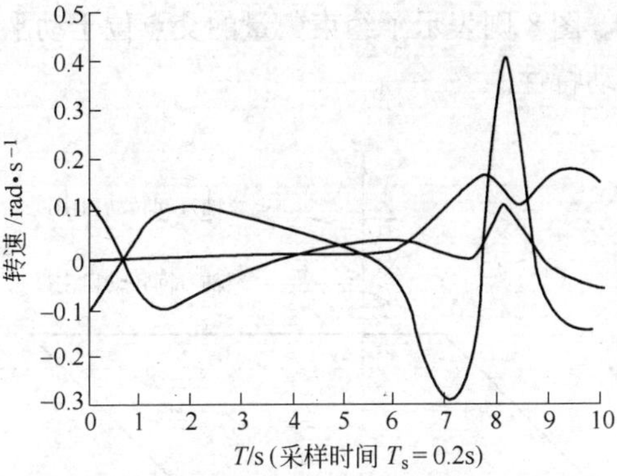

图 6　模型 2 中动平台的 3 个旋转运动

动是由于某一关联点的运动而导致的在某方向上的运动，其运动性质是由关联点所决定的。产生关联运动特性如图 7 所示。

模型 1 坐标系建立见图 1，绝对坐标系建立在定平台上，坐标原点 o 位于定平台中心，x、y 轴方向分别与定平台两相邻边平行，z 轴方向根据右手法则确定并指向动平台；相对坐标系的坐标原点位于各运动副轴线相交点 P，x'、y'坐标轴线方向分别与 x、y 轴方向一致，z'与 z 轴重合，且到动平台中心点 o'的距离为 h；动平台相对坐标系

$\{o'\}$ 建立原则与绝对坐标系相一致，点 o' 在坐标系 $\{P\}$ 中的初始位置用 ${}^{P}\boldsymbol{P}_{o'}$ 来表示，其形式为：${}^{P}\boldsymbol{P}_{o'}=(0,0,h)$。

设 o' 绕相对坐标系 $\{P\}$ 三个坐标轴的转动角分别为 α、β、γ 以及沿 z' 轴线方向的一个平移为 d，根据 D-H 方法可知，点 o' 在相对坐标系 $\{P\}$ 中经过变换后的位置用 ${}^{P}P_{o''}$ 来表示，其形式为：

$$
{}^{P}\boldsymbol{P}_{o''}={}^{P}_{o'}\boldsymbol{R}_{x'y'z'}\ (\alpha,\ \beta,\ \gamma)\,{}^{P}\boldsymbol{P}_{o'}+{}^{P}_{o'}\boldsymbol{T}\ (z',\ d) \tag{1}
$$

式中，${}^{P}_{o'}\boldsymbol{R}_{x'y'z'}(\alpha,\beta,\gamma)=R(z_P,\alpha)R(y_P,\beta)R(x_P,\gamma)$，${}^{P}_{o'}\boldsymbol{T}(z',d)=[0\quad 0\quad d\quad 1]^{\mathrm{T}}$

则经过变换后的动平台原点的坐标为：

$$
{}^{P}\boldsymbol{P}_{o''}=(-h\sin\beta,h\cos\beta\sin\gamma,h\cos\beta\cos\gamma+d) \tag{2}
$$

根据式（2）可知，动平台上的参考点（原点）的坐标由于相交点 P 的影响存在 3 个关联运动，此关联运动与交叉点到动平台原点的距离 h 有关。图 8 则表示了约束矢量的交点位于动平台上时，动平台所具有的运动特性。

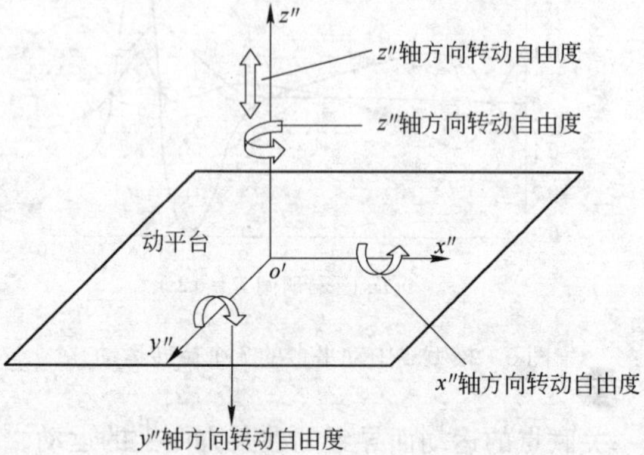

图 7　4-RRCR 型并联机构关联运动特性

1.3.4　刚体传感器配置在约束矢量交点时的仿真结果分析

为了说明点 P 的运动轨迹，我们将刚体传感器（Matlab SimMechanics 中所提供的 Body Sensor 模块，用于检测刚体上点的位移及转

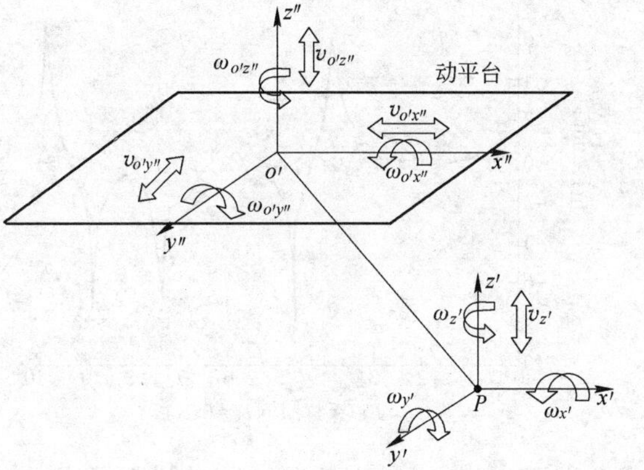

图 8　R、C 副轴线交点位于动平台时机构运动特性

动速度）放置在 R、C 副轴线交点处，分别在传感器中选择位移（mm）及转动速度（rad/s）作为信号输出。传感器的位姿信号输出如图 9 和图 10 所示。

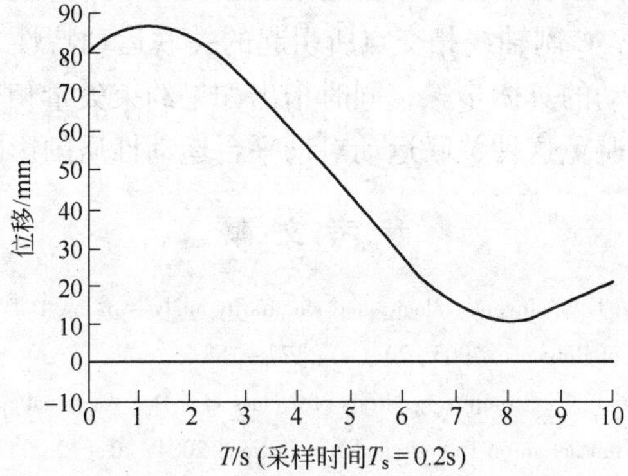

图 9　R、C 副轴线交点的一个平移运动

根据图 3 ~ 图 6 及图 9 与图 10 的对比可知，在设计 4-RRCR 型 3R1T 并联机构中，可以适当地选择第二组转动副轴线交点的位置，以期达到所设计动平台运动特性的要求。

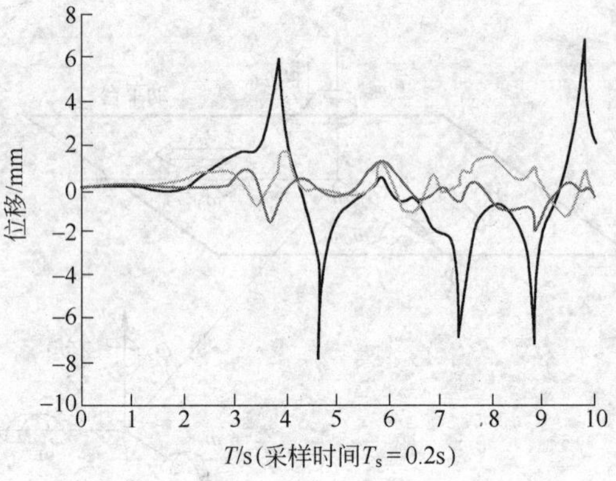

图 10　R、C 副轴线交点的 3 个旋转运动

1.3.5　结论

通过对 4-RRCR 型具有 3R1T 并联机构动平台运动性质的分析，揭示了在该类并联机构的设计中存在关联运动特性，而这种运动特性对所设计出的并联机器人动平台运动性质具有不可忽视的影响。本文分析了由于 R、C 副轴线相交点所引起的关联运动特性，并给出了计算这种关联运动的具体步骤，同时指出如果约束矢量相交点作用在动平台原点时可避免这种关联运动对动平台运动性质的影响。

参考文献

[1] Fang Y F, Tsai L W. Inverse velocity and singurarity analysis of low-DoF several manipulator [J]. Journal of Robotics, 2003, 20 (4): 177 ~ 188.

[2] Fang Y F, Tsai L W. Structure synthesis of a class of 3-DoF rotational parallel manipulators [J]. IEEE Transaction on Robotic and Automation, 2004, 20 (1): 117 ~ 121.

[3] Fang Y F, Tsai L W. Structure synthesis of 4-DoF and 5-DoF parallel manipulator with identical limbs [J]. International Journal of Robotics Research, 2002, 21 (9): 799 ~ 810.

[4] Li Q C, Huang Z. Type synthesis of 4-DoF parallel manipulator [C] //Proceedings of the 2005 IEEE International Conference on Robotics and Automation, Taipei, Taiwan, 2005, 755 ~ 761.

[5] Hunt K H. Kinematic Geometry of Mechanisms [M]. Oxford, Great-Britain: Oxford Univer-

sity Press, 1978.

[6] 黄真，孔令富，方跃法．并联机器人机构学理论与控制［M］．北京：机械工业出版社，1997.

[7] 程志红，谭建荣，段雄等．面向交互的平面连杆机构运动学分析的计算机实现［J］．中国机械工程，2004，15（7）：607～610.

1.4 基于螺旋理论的少自由度机器人速度反解及奇异性分析

本文采用螺旋理论对少自由度（自由度少于6）机器人的速度反解及奇异性进行了分析。根据机器人反螺旋系统确定了笛卡儿操作空间到关节空间的速度映射关系，并由螺旋及反螺旋理论所获得的雅克比矩阵对少自由度机器人的奇异性进行了分析。通过对四自由度 Stanford 型机器人的仿真研究说明了该方法的有效性。

1.4.1 引言

在螺旋及反螺旋理论的基础上，对少自由度机器人速度反解及奇异性进行了分析。通过将末端操作器6个运动参数分解为两个部分：线性相关部分和最大线性无关组部分，从而建立了少自由度操作空间与关节空间的速度映射关系。采用反螺旋理论确定组成最大线性无关组中的基本运动参数。

1.4.2 速度映射关系及奇异性条件

少自由度机器人操作空间速度 $\boldsymbol{v}$ 与关节空间速度 $\dot{\boldsymbol{q}}$ 的线性映射关系可通过速度雅克比矩阵 $\boldsymbol{J}$ 来表示：

$$\boldsymbol{v} = \boldsymbol{J}\dot{\boldsymbol{q}} \tag{1}$$

式中，$\boldsymbol{v} = [\boldsymbol{\omega}_x \quad \boldsymbol{\omega}_y \quad \boldsymbol{\omega}_z \quad \boldsymbol{v}_x \quad \boldsymbol{v}_y \quad \boldsymbol{v}_z]^{\mathrm{T}}$，为末端操作器六维速度矢量，包含三维方向转动角速度矢量及三维方向移动速度矢量；$\dot{\boldsymbol{q}} = [\dot{\boldsymbol{q}}_1, \dot{\boldsymbol{q}}_2, \cdots, \dot{\boldsymbol{q}}_n]^{\mathrm{T}}$，为少自由度机器人（$n$ 自由度）关节空间中 n 个关节速度矢量形式；$\boldsymbol{J}$ 为少自由度机器人速度雅克比矩阵，$\boldsymbol{J} \in \mathbf{R}^{6 \times n}$。对于少自由度机器人而言，由于 $n < 6$，所以从关节空间到操作空间的速度反解映射（即雅克比矩阵 $\boldsymbol{J}$ 的逆矩阵）无法求解，通

常采用伪逆矩阵形式求解，形式为：

$$\dot{\boldsymbol{q}} = \boldsymbol{J}^{+}\boldsymbol{v} \tag{2}$$

式中，$\boldsymbol{J}^{+} = (\boldsymbol{J}^{\mathrm{T}}\boldsymbol{J})^{-1}\boldsymbol{J}^{\mathrm{T}}$ 为伪逆矩阵形式，因此奇异性条件可定义为：

$$\det(\boldsymbol{J}^{\mathrm{T}}\boldsymbol{J}) = 0 \tag{3}$$

在少自由度机器人中，由于 $\boldsymbol{J}^{+}$ 是 $\boldsymbol{J}$ 逆的近似解，不能准确描述关节空间对操作空间的真实映射关系，对于 n（$n<6$）自由度机器人，在操作空间中只有 n 个独立速度矢量与关节空间 n 个关节速度矢量形成一一映射关系，而其余 $6-n$ 个操作空间速度矢量则与此关节速度矢量线性相关。将操作空间中 n 个独立速度矢量表示为包含独立参数的 $\boldsymbol{v}_1 \in \mathbf{R}^{n}$ 和与独立速度矢量 $\boldsymbol{v}_2 \in \mathbf{R}^{6-n}$ 线性相关的速度矢量，并且 $\boldsymbol{v}_2$ 可以用 $\boldsymbol{v}_1$ 的函数形式表示为：

$$\boldsymbol{v}_2 = f(\boldsymbol{v}_1) \tag{4}$$

通过以上定义，式（1）可以表示为：

$$\begin{bmatrix} \boldsymbol{v}_1 \\ \vdots \\ \boldsymbol{v}_2 \end{bmatrix}_{n \times 1} = \begin{bmatrix} \boldsymbol{J}_1 \\ \vdots \\ \boldsymbol{J}_2 \end{bmatrix}_{6 \times n} \dot{\boldsymbol{q}} \tag{5}$$

式中，$\boldsymbol{J}_1 \in \mathbf{R}^{n \times n}$；$\boldsymbol{J}_2 \in \mathbf{R}^{(6-n) \times n}$，$\boldsymbol{J}_2$ 中的每一行元素与 $\boldsymbol{J}_1$ 中的每一行元素线性相关。上式又可写为以下形式：

$$\boldsymbol{v}_1 = \boldsymbol{J}_1 \dot{\boldsymbol{q}} \tag{6}$$

当少自由度机器人处于非奇异位形时，$\boldsymbol{J}_1$ 具有逆矩阵，则有：

$$\dot{\boldsymbol{q}} = \boldsymbol{J}_1^{-1} \boldsymbol{v}_1 \tag{7}$$

式（6）、式（7）表示了从操作空间到关节空间速度一一映射关系，本文将采用螺旋与反螺旋理论寻求一种有效的确定速度映射的雅克比矩阵并将操作空间速度进行分解的方法。

从式（6）可知，对于少自由度机器人奇异性问题，可采用如下定义形式：

$$\det(\boldsymbol{J}_1) = 0 \tag{8}$$

1.4.3 操作空间独立运动矢量的确定

1.4.3.1 关节螺旋系及其反螺旋形式

一单位螺旋可定义为 6×1 矩阵形式：

$$\hat{\$}_j = \begin{bmatrix} s_j \\ r_j \times s_j + \lambda s_j \end{bmatrix} \tag{9}$$

式中，s_1 螺旋轴线方向的单位矢量；r_1 关节轴线上任意点在参考坐标系中的位置矢量；λ 沿螺旋方向上的节距；对于转动副，$\lambda=0$；对于移动副，$\lambda=\infty$。下标用于表示第 j 个关节的螺旋系。

机器人末端操作器的瞬时运动可以通过各关节轴线瞬时运动的线性组合形式来表示：

$$\$_n = [\hat{\$}_1, \hat{\$}_2, \cdots, \hat{\$}_n] = \begin{bmatrix} \dot{q}_1 \\ \dot{q}_2 \\ \vdots \\ \dot{q}_n \end{bmatrix} \tag{10}$$

式中，$\hat{\$}_n = \begin{bmatrix} \omega_n \\ v_n \end{bmatrix}$；$\omega_n = [\omega_x \quad \omega_y \quad \omega_z]^T$ 机器人末端操作器角速度矢量；$v_n = [v_x \quad v_y \quad v_z]^T$ 机器人末端操作器线速度矢量；$\dot{q}_i$（$i=1$，2，…，n）第 i 个关节的关节速度矢量。

从式（10）可知，机器人的雅克比矩阵可用各关节的单位螺旋形式来表示：

$$J = [\hat{\$}_1, \hat{\$}_2, \cdots, \hat{\$}_n] \tag{11}$$

对于少自由度机器人系统，运动螺旋系 $n<6$，根据螺旋与反螺旋理论，该运动螺旋系存在 $6-n$ 个反螺旋，与运动螺旋相对应的反螺旋称为力螺旋，力螺旋定义为：

$$\$_r = \begin{bmatrix} f \\ r_0 \times f + c \end{bmatrix} = \rho\ \hat{\$}_r \tag{12}$$

式中，f 和 c 为作用在位置矢量 r_0 上的力和力偶；ρ 为单位力螺旋的放大系数。

单位力螺旋定义为：

$$\hat{\$}_r = \begin{bmatrix} s_r \\ r_0 \times s_r + \lambda_r s_r \end{bmatrix} \tag{13}$$

式中，s_r 沿力作用方向上的单位矢量形式；λ_r 单位力螺旋节距：当力螺旋表示为纯力形式时，$\lambda_r=0$；当力螺旋表示为纯力偶形式时，$\lambda_r=\infty$。

所求出的 $6-n$ 阶反螺旋与单位关节螺旋的互易积为：

$$\hat{\$}_j^{\mathrm{T}}\,\$_{ri}=0\ (j=1,2,\cdots,6-n;i=1,2,\cdots,n) \tag{14}$$

1.4.3.2 零空间方法确定反螺旋形式

对于 n 阶少自由度机器人系统，为了求解方程，需要确定 $6-n$ 个反螺旋形式。本节采用零空间方法来确定反螺旋形式。

将式（14）经过 $j=1$，2，…，n 次方程迭代，写成矩阵形式为：

$$\boldsymbol{J}^*\$_{ri}=[0]\quad i=1,2,\cdots,6-n \tag{15}$$

式中，$\boldsymbol{J}^*\$_{ri}=\boldsymbol{\mu J}$；$\mu=\begin{bmatrix}0_{3\times3} & \cdots & I_{3\times3}\\ \vdots & & \vdots\\ I_{3\times3} & \cdots & 0_{3\times3}\end{bmatrix}$。

将 $\boldsymbol{J}^*$ 的零空间表示为 $N(\boldsymbol{J}^*)$，则式（15）表示了 $\$_{ri}=\boldsymbol{N}(\boldsymbol{J}^*)$，也就是说所求出的 $6-n$ 个反螺旋为 $\boldsymbol{N}(\boldsymbol{J}^*)$ 的最大线性无关组，因此求解 $6-n$ 个反螺旋的实质就是求出 $\boldsymbol{J}^*$ 的零空间解的形式。

设 $\boldsymbol{u}$ 为六维矢量，则 $\boldsymbol{u}$ 与其零空间矩阵 $\boldsymbol{J}^*$ 满足下列关系：

$$\boldsymbol{J}^*\boldsymbol{u}=[0] \tag{16}$$

式（16）对于矢量 $\boldsymbol{u}$ 中的 6 个未知参数包含了 n 个线性方程，由于 $n<6$，所以所表示的方程组为非确定系统。设 $\boldsymbol{J}^*$ 矩阵的秩为满秩 n，将矩阵 $\boldsymbol{J}^*$ 和矢量进行如下形式的分解：

$$\boldsymbol{J}^*=[\boldsymbol{J}_p\ \ \vdots\ \ \boldsymbol{J}_s],\ \boldsymbol{u}=\begin{bmatrix}\boldsymbol{u}_p\\ \boldsymbol{u}_s\end{bmatrix} \tag{17}$$

式中，$\boldsymbol{J}_p\in\mathbf{R}^{n\times n}$，$\boldsymbol{J}_s\in\mathbf{R}^{n\times(6-n)}$，$\boldsymbol{u}_p=[\boldsymbol{u}_1\ \ \boldsymbol{u}_2\ \ \cdots\ \ \boldsymbol{u}_n]^{\mathrm{T}}$，$\boldsymbol{u}_s=[\boldsymbol{u}_{n+1}\ \ \boldsymbol{u}_{n+2}\ \ \cdots\ \ \boldsymbol{u}_6]^{\mathrm{T}}$。

对于所选定的 $\boldsymbol{u}_s$，根据 Cramer 准则，有唯一的与之相对应的 $\boldsymbol{u}_p$。由于 $\boldsymbol{u}_s$ 为 $(6-n)\times1$ 维向量，所以对于零空间 $\boldsymbol{J}^*$ 存在 $6-n$ 个解。任一属于零空间解矢量 $\hat{\boldsymbol{u}}_i$（$i=1$，2，…，$6-n$）都可以用一组最大线性无关矢量的线性组合来表示，其形式为：

$$\boldsymbol{u}=k_1\hat{\boldsymbol{u}}_1+k_2\hat{\boldsymbol{u}}_2+\cdots+k_{6-n}\hat{\boldsymbol{u}}_{6-n} \tag{18}$$

式中，k_i 为定常系数。

每个基向量中的六个元素可表示为：$\hat{\boldsymbol{u}}_j = [u_{j1} \quad u_{j2} \quad u_{j3} \quad u_{j4} \quad u_{j5} \quad u_{j6}]^{\mathrm{T}} (j = 1,2,\cdots,6-n)$

有下列定义：

$$\boldsymbol{u}_{jk} = \begin{cases} -\dfrac{\det(\boldsymbol{J}_{kj}^{*})}{\det(\boldsymbol{J}_{p}^{*})} & k = 1,2,\cdots,n; k = j \\ 1 & k > j \\ 0 & k < j \end{cases} \tag{19}$$

式中，$\boldsymbol{J}_{kj}^{*}$ 由 $\boldsymbol{J}_p$ 矩阵中的第 k 行与 $\boldsymbol{J}_s$ 矩阵中的第 j 行所组成。

通过对所构成的少自由度机器人雅克比矩阵分析，可以得到少自由度机器人产生奇异性的条件。

1.4.4 Stanford 型机器人奇异性分析

四自由度 Stanford 型机器人各关节坐标系建立如图 1 所示。

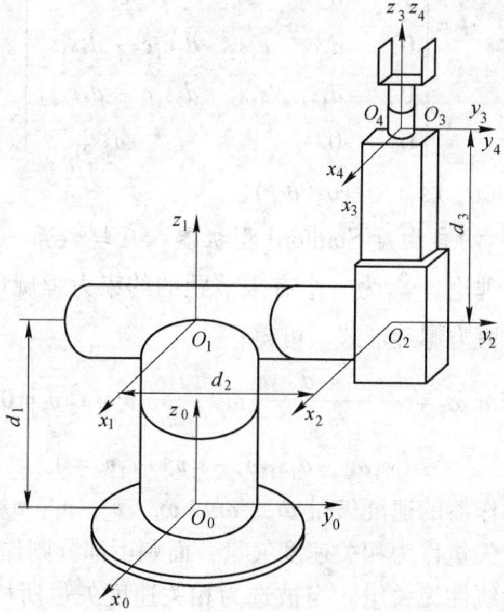

图 1　四自由度 Stanford 型机器人结构示意图

四自由度 Stanford 型机器人由 3 个转动副和 1 个移动副所组成，

相应的 D－H 参数如表 1 所示。

表 1 四自由度 Stanford 型机器人 D－H 参数

关 节	θ_i	α_i	a_i	d_i
1	θ_1	0^0	0	d_1
2	θ_2	$-\pi/2$	0	d_2
3	0	$\pi/2$	0	d_3
4	θ_4	0^0	0	0

1.4.4.1 四自由度 Stanford 型机器人雅克比矩阵

四自由度 Stanford 型机器人固定坐标系选取在基座上，标记为 $\{o_0-x_0y_0z_0\}$，该类型机器人雅克比矩阵为 6×4 矩阵形式：

$$
{}^0\boldsymbol{J}=\begin{bmatrix} 0 & -s_1 & 0 & c_1s_2 \\ 0 & c_1 & 0 & s_1s_2 \\ 1 & 0 & 0 & c_2 \\ 0 & -d_1c_1 & c_1s_2 & d_2c_1c_2-d_1s_1s_2 \\ 0 & -d_1s_1 & s_1s_2 & d_2s_1c_2+d_1c_1s_2 \\ 0 & 0 & c_2 & -d_2s_2 \end{bmatrix} \tag{20}
$$

式中，$s_*=\sin(\theta_*)$，$c_*=\cos(\theta_*)$。

1.4.4.2 四自由度 Stanford 型机器人反螺旋系

根据螺旋理论，$\hat{\$}_{r1}$ 为一个有限节距的约束力螺旋形式；$\hat{\$}_{r2}$ 为一个零节距的约束力螺旋形式。可得：

$$
\begin{gathered} d_2c_1\boldsymbol{\omega}_x+\frac{d_2s_1c_1-d_1s_2c_2}{c_1}\boldsymbol{\omega}_y-\frac{s_2c_2}{c_1}\boldsymbol{v}_x+s_2^2\boldsymbol{v}_z=0, \\ -d_1c_1\boldsymbol{\omega}_x-d_1s_1\boldsymbol{\omega}_y-s_1\boldsymbol{v}_x+c_1\boldsymbol{v}_y=0 \end{gathered} \tag{21}
$$

在末端操作器的速度矢量 $\boldsymbol{\omega}_x$，$\boldsymbol{\omega}_y$，$\boldsymbol{\omega}_z$，$\boldsymbol{v}_x$，$\boldsymbol{v}_y$，$\boldsymbol{v}_z$ 中，可以选择任意两个速度矢量作为相关速度矢量，而剩下部分则作为独立的速度矢量，即最大线性无关组。当被选为相关速度矢量所构成的 2×2 系数矩阵行列式等于零时，在此条件下机器人产生奇异。由于 $\boldsymbol{\omega}_z$ 在式（21）中未出现而不能被选为相关速度矢量，所以按此方法从剩余 5 个矢量中选取 2 个相关速度矢量有 10 种可能的组合方式。

1.4.4.3 奇异性分析

选择 $\boldsymbol{\omega}_x$ 和 $\boldsymbol{\omega}_y$ 作为相关速度矢量，根据式（21）可知，组成 $\boldsymbol{\omega}_x$ 的 $\boldsymbol{\omega}_y$ 阶系数矩阵形式为：

$$\boldsymbol{M}_{12}=\begin{bmatrix} d_2c_1 & \dfrac{d_2s_1c_1-d_1s_2c_2}{c_1} \\ -d_1c_1 & -d_1s_1 \end{bmatrix} \tag{22}$$

其矩阵行列式为：$\det(\boldsymbol{M}_{12})=-d_1^2s_2c_2$

雅克比矩阵 $\boldsymbol{J}_1$ 的行列式为：$\det(\boldsymbol{J}_1)=d_1^2s_2c_2$

设 $\det(\boldsymbol{J}_1)=0$，则少自由度机器人在此假设条件下产生运动的奇异条件为：

（1）$\sin\theta_2=0$；

（2）$\cos\theta_2=0$。

1.4.5 仿真研究

在仿真研究中，设 $d_1=300\text{mm}$，$d_2=200\text{mm}$。给定的跟踪轨迹为：$y=200\text{mm}$，$z=0\text{mm}$。

设坐标 $\{o_4-x_4y_4z_4\}$ 原点为操作空间中轨迹跟踪点。机器人初始运行条件为：$\boldsymbol{q}=[0°\quad 120°\quad 600\text{mm}\quad 0°]^{\mathrm{T}}$，$\dot{\boldsymbol{q}}=[0\quad 0\quad 0\quad 0]^{\mathrm{T}}$。

对于末端操作器速度最大线性无关组的选取为：${}^0\boldsymbol{v}=[\boldsymbol{\omega}_z\quad \boldsymbol{v}_x\quad \boldsymbol{v}_y\quad \boldsymbol{v}_z]^{\mathrm{T}}$，其中，$\boldsymbol{\omega}_z=\boldsymbol{v}_z=\boldsymbol{v}_y=0$，$\boldsymbol{v}_x=\boldsymbol{v}_{\max}\sin\left(\dfrac{\pi t}{T}\right)\text{mm/s}$。

仿真曲线如图 2 ~ 图 5 所示。

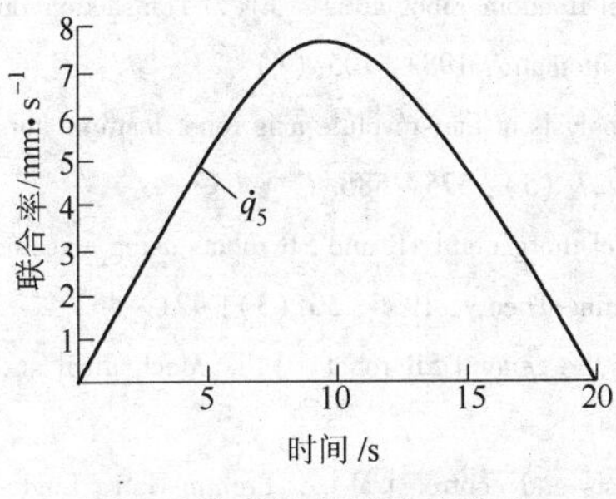

图 2　移动关节线速度仿真

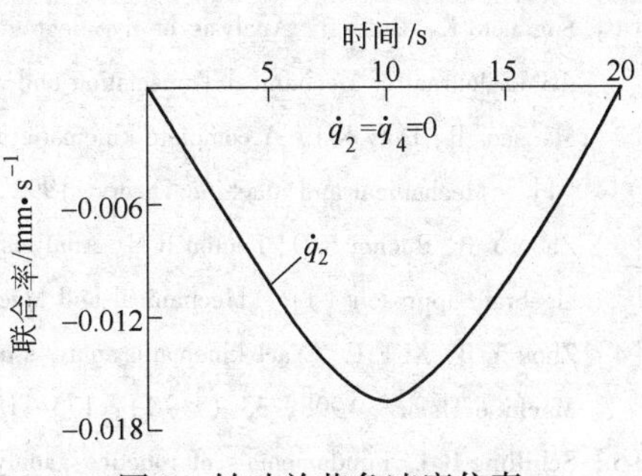

图 3　转动关节角速度仿真

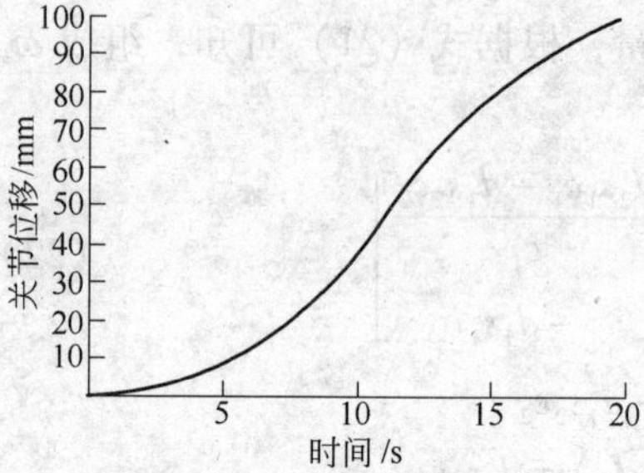

图 4 四自由度 Stanford 型机器人移动关节位置仿真

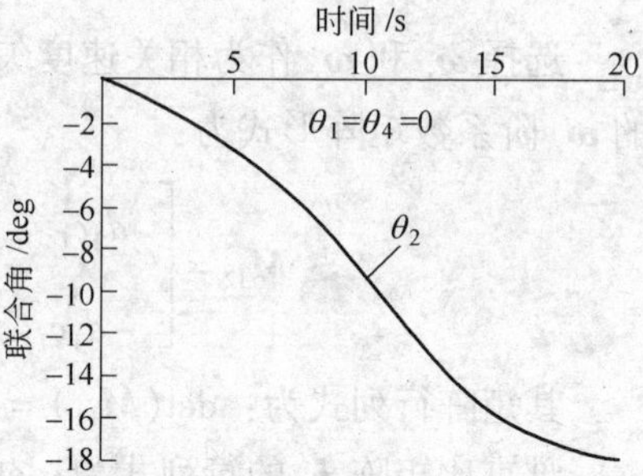

图 5 四自由度 Stanford 型机器人转动关节位置仿真

1.4.6 结论

在螺旋及反螺旋理论的基础上，本文对少自由度机器人速度反解及奇异性进行了分析。通过将末端操作器 6 个运动参数分解为两个部分：线性相关部分和最大线性无关组部分，从而建立了少自由度操作空间与关节空间的速度一一映射关系。采用反螺旋理论确定组成最大线性无关组中的基本运动参数。通过对四自由度 Stanford 型机器人的关节位置及速度的仿真研究说明了该方法的有效性，从而为少自由度机器人的速度反解提供了一条可行的研究途径。

参 考 文 献

[1] Sugimoto K, Duffy J. Analysis of five-degree-of-freedom robot arms [J]. Transaction on ASME Journal of Mechanical Transmation and Automatic, 1983, 105 (1): 23 ~ 27.

[2] Manseur R, Doty K L. A complete kinematic analysis of four-revolute-axis robot manipulators [J]. Mechanical and Machine Theory, 1992, 27 (5): 575 ~ 586.

[3] Zhou Y B, Buchal R O, Fenton R G. Analysis of the general 4R and 5R robots using a vector algebraic approach [J]. Mechanical and Machine Theory, 1995, 30 (3): 421 ~ 432.

[4] Zhou Y B, Xi F F. Exact kinematic analysis of the general 5R robot [J]. Mechanical and Machine Theory, 1998, 33 (1 - 2): 175 ~ 184.

[5] Schilling R J. Fundamentals of robotics: analysis and control [M]. Prentic-Hall, Englewoods, 1990.

[6] Wang H B, Ishimatsu T, Schaerer C, Huang Z. Kinematics of a five degree-of-freedom prosthetic arm [J]. Mechanical and Machine Theory, 1998, 33 (7): 895 ~ 908.

1.5 基于螺旋理论的3-RPS型并联机器人运动学分析

3-RPS型并联机构具有三个结构对称的支链形式，各支链由一个转动副与基座相连接，一个球面副与动平台相连接，转动副与球面副由移动副所连接。本文采用螺旋理论及空间机构构型原理，通过约束形式分析得出该类型并联机器人运动性质，采用矢量分析方法对其运动学正解/反解进行求解，得出该并联机构运动学方程。基于对称非齐次少自由度并联机构Jacobine矩阵，进一步对该类型并联机器人结构奇异性进行分析与总结。

1.5.1 引言

少自由度并联机构越来越成为研究者所关注的一个新的研究领域。少自由度并联机构在某些领域的成功应用显示出其卓越的结构性能[1~6]。科研人员采用反螺旋理论对一类具有三自由度转动并联机器人支链结构进行了分析和综合，归纳出组成三自由度纯转动并联机器人所具备的可能的支链形式[7]。一些科研人员列举了一类具有四支链的三自由度并联机构并提出该类型并联机构运动学的分析方法。Miller[9]基于螺旋理论提出对具有三自由度纯移动并联机构的分析方法，并分析了该类机构驱动关节奇异性条件。本文采用螺旋理论对具有3-RPS支链形式的三自由度并联机构进行运动学特性分析，并基于矢量分析方法求解该类型并联机器人位置正/反解，通过对Jocabian矩阵秩的分析得出该类型并联机构奇异性条件。

1.5.2 3-RPS型并联机器人运动学分析

1.5.2.1 3-RPS型并联机器人位置分析

静平台的惯性坐标系由 $\{o-xyz\}$ 表示，同时建立动平台惯性坐标系，由 $\{o'-uvw\}$ 表示。为了简化运动学模型，基座惯性坐标系的 x 轴线方向与 A_1A_2 连线方向一致，y 轴与 x 轴线正交，z 轴由右手螺旋法则确定。

坐标原点 o 位于静平台中心。动平台惯性坐标的确定方法与静平台一致。$\boldsymbol{p}=[x \quad y \quad z]^{\mathrm{T}}$ 为从基坐标系原点 o 指向动坐标系原点 o' 的矢量；$\boldsymbol{r}_i=[r_{ix} \quad r_{iy} \quad r_{iz}]^{\mathrm{T}}$ 为从基坐标系原点 o 指向与动平台相连接的第 i 个关节 B_i 的矢量；$\boldsymbol{a}_i=[a_{ix} \quad a_{iy} \quad a_{iz}]^{\mathrm{T}}$ 为从基坐标系原点 o 指向与静基座相连接的第 i 个关节 A_i 的矢量；$\boldsymbol{b}_i=[b_{ix} \quad b_{iy} \quad b_{iz}]^{\mathrm{T}}$ 为从动坐标系原点 o' 指向与动平台相连接的第 i 个关节 B_i 的矢量；$\boldsymbol{l}_i=[l_{ix} \quad l_{iy} \quad l_{iz}]^{\mathrm{T}}$ 为从与基座相连接的第 i 个关节 A_i 指向与动平台相连接的第 i 个关节 B_i 的矢量；$\|l_i\|$对于各支链而言为固定参数。

3-RPS 型并联机器人结构如图 1 所示。

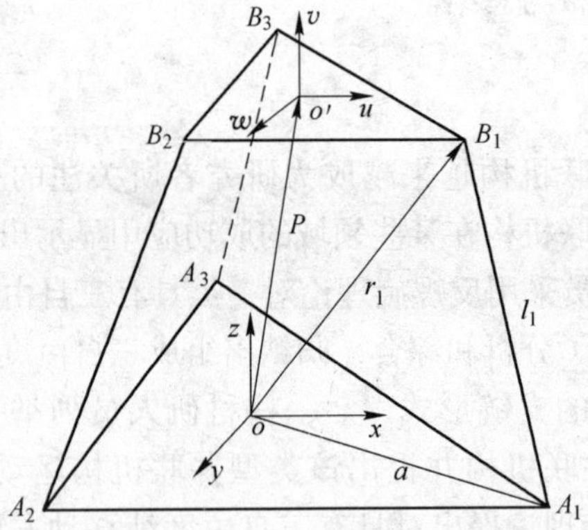

图 1　3-RPS 型并联机器人结构示意图

由图 1 可知，各支链的环路矢量方程为：

$$l_i\boldsymbol{s}_i=\boldsymbol{p}+\boldsymbol{b}_i-\boldsymbol{a}_i \tag{1}$$

式中，l_i 各支链的长度，标量值；$\boldsymbol{s}_i$ 支链的矢量方向，矢量值。

1.5.2.2　3-RPS 型并联机器人位置反解

对于机构的反解，其位置矢量为给定值，解的结果为驱动参数，该参数可以由转动关节的角度来表示（驱动类型为转动），也可以由连杆的长度参数来表示（直线驱动）。

三个连杆长度 d_i（$i=1，2，3$）可以由下式求解：

$$d_i=\pm\sqrt{(\boldsymbol{p}+{}^{A}\boldsymbol{b}_i-\boldsymbol{a}_i)^{\mathrm{T}}(\boldsymbol{p}+{}^{A}\boldsymbol{b}_i-\boldsymbol{a}_i)} \tag{2}$$

从求解的过程来看，对应于每个末端执行器所提供的位置矢量，每个连杆有两个可能的位置解，经组合可知，该类型并联机构存在8组位置反解形式。

1.5.2.3 3-RPS 型并联机器人位置正解

对于3-RPS型并联机构的正解，由于$^{A}\boldsymbol{B}_i$可通过D－H方法求出，因此连杆的长度d_i（$i=1$，2，3）可以由$d_i=\|\boldsymbol{A}_i{}^{A}\boldsymbol{B}_i\|\boldsymbol{d}_i$所求出，矢量方程满足以下条件：

$$\boldsymbol{p}^2-2p^{\mathrm{T}}(\boldsymbol{a}_i-\boldsymbol{b}_i)+(\boldsymbol{a}_i-\boldsymbol{b}_i)^2=\boldsymbol{d}_i^2 \tag{3}$$

1.5.3 3-RPS型并联机器人奇异性分析

当并联机构处于奇异位形时，机构将失去或得到一个或多个自由度，机构的运动学特性将随之改变，在奇异点处机构的加速度将大幅增加，刚度变差。

因此，并联机构的奇异位形及奇异空间分析是并联机器人设计阶段的一个重要组成部分。

1.5.3.1 3-RPS 型并联机器人雅克比矩阵分析

3-RPS型并联机器人运动学支链上的速度矢量形式可以表示为：

$$\boldsymbol{v}_p=\$_1\theta_1+\$_2 d_2+\$_3\theta_3+\$_4\theta_4+\$_5\theta_5 \tag{4}$$

雅克比矩阵形式为：

$$\boldsymbol{M}_{c_1}=\begin{bmatrix}\hat{\$}_{r1}\\ \hat{\$}_{r2}\\ \hat{\$}_{r3}\end{bmatrix}\begin{bmatrix}0 & \boldsymbol{I}_{3\times3}\\ \boldsymbol{I}_{3\times3} & 0\end{bmatrix},\ \boldsymbol{M}_{c_2}=\frac{\hat{\$}_{ri}\begin{bmatrix}0 & \boldsymbol{I}_{3\times3}\\ \boldsymbol{I}_{3\times3} & 0\end{bmatrix}\hat{\$}_{i1}}{\hat{\$}_{ri}\begin{bmatrix}0 & \boldsymbol{I}_{3\times3}\\ \boldsymbol{I}_{3\times3} & 0\end{bmatrix}} \tag{5}$$

当$\hat{\$}_{ri}\begin{bmatrix}0 & \boldsymbol{I}_{3\times3}\\ \boldsymbol{I}_{3\times3} & 0\end{bmatrix}$的秩小于3，称之为配置奇异；

当$\hat{\$}_{ri}\begin{bmatrix}0 & \boldsymbol{I}_{3\times3}\\ \boldsymbol{I}_{3\times3} & 0\end{bmatrix}\hat{\$}_{i1}$的秩小于3，称之为主动奇异；

当两种形式均发生时，称之为机构奇异。

1.5.3.2 3-RPS 型并联机器人结构奇异位形分析

3支链上的反螺旋所组成的雅克比矩阵形式为：

$$
\boldsymbol{M}_{c_1} = \begin{bmatrix} 0 & l_1 s\alpha_1 & l_1 c\alpha_1 - \dfrac{\sqrt{3}}{3}a & 1 & 0 & 0 \\ N_1 & M_1 & \dfrac{ac\alpha_2^2 + as\alpha_2 - \dfrac{4\sqrt{3}}{3}l_2}{6} & \dfrac{\sqrt{3}}{3} & 1 & 0 \\ -l_3 s\alpha_3 & -\dfrac{l_3 s\alpha_3}{\sqrt{3}} & \dfrac{2}{3}(\sqrt{3}l_3 c\alpha_3 - a) & -\dfrac{\sqrt{3}}{3} & 1 & 0 \end{bmatrix} \tag{6}
$$

式中，$s\alpha_i = \sin\alpha_i$，$c\alpha_i = \cos\alpha_i$。

当 M_{c_1} 的秩小于 3 则发生结构奇异位形，由此可以得到结构奇异位形发生的条件。

1.5.4 结论

分析了 3-RPS 型并联机器人的运动性质，通过采用螺旋理论分析了各支链所组成的螺旋系得到其雅克比矩阵，并在此基础上分析了 3-RPS 型并联机器人奇异位形产生的条件，为今后此类并联机构的实际应用奠定理论基础。

参 考 文 献

[1] Lee K, Shah D. Kinematic analysis of a three degree - of - freedom in parallel actuated manipulator [J]. IEEE Journal of Robotics and Automation, 1988, 4 (3): 354 ~360.

[2] Yang P H, Waldron K J, Orin D E. Kinematics of a three degree - of - freedom motion platform for a low cost driving simulator [C] //Proceedings of the 5th International Symposium of Advances in Robot Kinematics, 1996, 89 ~98.

[3] Tsai L W, Walsh G C, Stamper R. Kinematics of a novel three degree - of - freedom translational platform [C] //Proceedings of the IEEE International Conference on Robotics and Automation, 1996, 3446 ~3451.

[4] Gosselin C M, Angeles J. The optimum kinematic design of planar three degree - of - freedom parallel manipulator [J]. ASME Journal of Mechanical, Transmissions and Automation Design, 1988, 110 (1): 35 ~41.

[5] Carretero J A, Nahon M, Podhorodeski R P. Workspace analysis of a three degree-of-freedom parallel mechanism [C] //Proceedings of the 1998 IROS Conference, 1998, 1012 ~1026.

[6] Fang Y F, Tsai L W. Structure synthesis of a class of 3-DoF rotational parallel manipulator [J]. IEEE Robotics and Automation, 2004, 20 (1): 375 ~381.

[7] Sameer A J, Tsai L W. The kinematics of a class of 3-DoF, 4 – legged parallel manipulators [J]. Transactions of the ASME, 2003, 125 ~ 127.

[8] Kong X W, Gosselin C M. Type synthesis of 3-DoF translational parallel manipulators based on screw theory [J]. Journal of Mechanical Design, 2003, 126: 83 ~ 93.

[9] Miller K. Maximization of workspace volume of 3-DoF spatial parallel manipulators [J]. Journal of Mechanical Design, 2002, 124: 347 ~ 358.

1.6 Analytical Identification of Limb Structures at Special Displacement for Parallel Manipulator

A systematic analytical method, based on the theory of screws, is presented for identification of limb structures with a special displacement of general and over-constrained 3-degree-of-freedom (DoF) parallel manipulators. Given a system of wrenches of constraint acting upon a special displacement, the consentaneous basis screws of limbs are determined via the reciprocal screw and null-space theory. Then, the structures and displacement of the joints on the limb are obtained by linear combination of these basis screws. Feasible limbs can be used for construction of translational and rotational platform according to the constraint acting on a special displacement. The kinematical property of platform is presented.

1.6.1 Introduction

Parallel manipulator have attracted much attention among researchers after Stewart[1] proposed the first parallel manipulator platform, then Hunt[2] developed the screw theory to resolve the parallel manipulator structures. From literatures[3~7], the limb structures with the pure translational and rotational have included. However, when the limbs moving, there also have many special displacement these literatures have not dealt with.

The structural synthesis of low-DoF parallel manipulator involves the complicated constraint system. Each limb of a parallel manipulator applies one or more (less than 3) wrenches constraint to the moving platform, the possible mobility to the platform will be determined by screw system conbinated with limbs. Hence, the main task for analytical identification of limb structures at special displacement is to identify the joints arrangement of limbs. The early design method for such purpose based on experiences. Herve and Sparacino[8] presented the synthesis of new parallel structure based on the mathematical group theory and a family of 3 DoF robots for pure spatial translation movement has conceived. Frisoli et al[9] identified translational parallel actuated mechanism by screw algebra. Kong and Gosselin[10] analyzed the generation of parallel manipulators with three translational DoF based on screw theory. All of these literatures can not analysis the displacement and kinematics of parallel manipulators systemic. Fang systemic analyzed the limb structure of parallel manipulators and identified the joints arrangement on the limbs by using screw theory, the conceivable structure of limbs has been given in literatures[11~13].

In this paper, screw theory is used to analysis limb structure for 3-DoF parallel manipulators. The main contribution of this paper is consummate the procedure of limb joint screw system and explore a better completely list of the limb structure with the special displacement for 3-DoF parallel manipulators. There have two cases we considered: firstly, the reciprocal screw system of each limb restricts the platform rotation, as a result, the platform moving with translation; secondly, we restrict the platform with translational motions. When given the limb wrenches system, the joint screws are determined by a linear combination of the basis screws, the characteristics with the special displacement of the moving platform will be enumerated and classified according to the screws system of joints and links making up the limbs. The geometrical conditions for special displacement are also presented.

1.6.2 Reciprocal Screw System

The earliest references to screws can be traced back to the beginning of 19th century when Poinsot (1806) established the concept that and system of forces and a couple. Then Chaslest (1832) stated that rigid body displacement can be conceived as a rotation about a line accomplished by a translation along that line. This type of representation is canonical in form: moreover, it is subject to easy geometrical interpretation. Ball (1900) established a firm physical base for the mathematics of screws when he wrote his treatise on the small oscillations of a rigid body. The two fundamental concepts in the theory of screws is that the instantaneous motion of a rigid body is a twisting motion about the instantaneous screw axis, and that a system of forces acting upon the rigid body is a wrench acting about a particular screw axis.

In this section, the definition of screw and reciprocal screw are reviewed. Clifford proposed the allelomorph in 1873. The allelomorph is composed with two parts:

$$\hat{a} = a + \in \alpha_0 \tag{1}$$

where a is an original part, it called the Original allelomorph, then α_0 is called the allelomorph. Based on this theory, we can denote a movement by v and w, which is called twist. If denote a force and a couple, it is called wrench. We will describe the universality of screw as following.

When screw denoted by the vector, it is also can writing by six screw coordinates:

$$\hat{\$} = \begin{bmatrix} \boldsymbol{s} \\ \boldsymbol{s}_0 \end{bmatrix} \tag{2}$$

where $\boldsymbol{s}$ and $\boldsymbol{s}_0$ equal to a and α_0 in Eq. (1), it's geometry relationship can detailed as Fig. 1. where $\boldsymbol{s}$ is a vector pointing in the direction of screw axis, $\boldsymbol{s}_0$ is denoted by hypo-couple number. When screw is a linear vector, $\boldsymbol{s}_0 = \boldsymbol{r} \times \boldsymbol{s}$, $\boldsymbol{r}$ is the position vector of coordination point of any point on the screw axis. When $\boldsymbol{s}$ and $\boldsymbol{s}_0$ is not reciprocal, the pitch h is not equal to zero.

$$h = \frac{\boldsymbol{s} \cdot \boldsymbol{s}_0}{\boldsymbol{s} \cdot \boldsymbol{s}} \tag{3}$$

where h is called screw pitch, then Eq. (2) can denote by following.

$$\hat{\$} = \begin{bmatrix} \boldsymbol{s} \\ \boldsymbol{s}_0 + h\boldsymbol{s} \end{bmatrix} \tag{4}$$

If the pitch of a screw is infinite, the unit screw is defined as:

$$\hat{\$} = \begin{bmatrix} 0 \\ \boldsymbol{s} \end{bmatrix} \tag{5}$$

If the pitch of a screw is zero, the unit screw is defined as:

$$\hat{\$} = \begin{bmatrix} \boldsymbol{s} \\ \boldsymbol{r} \times \boldsymbol{s} \end{bmatrix} \tag{6}$$

The unit screw associated with a revolute joint is a screw of zero pitch pointing along the joint axis. The unit screw associated with a prismatic joint is a screw of infinite pitch pointing the direction of the joint axis.

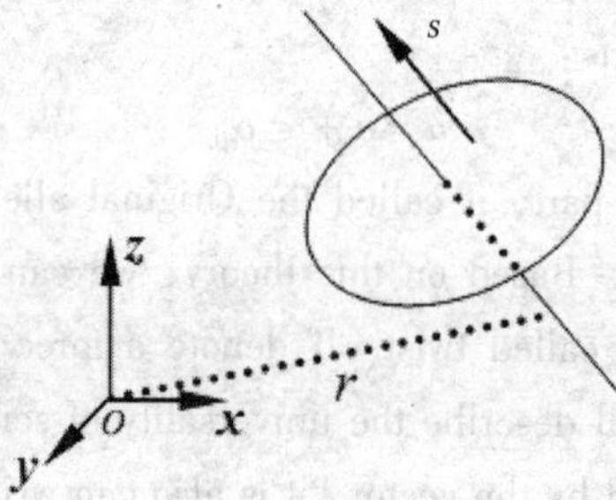

Fig. 1 Geometrical representation of screw

A rigid body, if there have no any restrict acting on it, has six DoF, including three rotational DoF and three translational DoF. If there have one couple constrain acting on the moving platform, on rotational DoF will be restricted, on the other hand, if there have one force constrain acting on the moving platform, one translational DoF will be restricted.

Two screws, $\$$ and $\$_r$, are said to reciprocal each other if they satisfy the condition:

$$\$ \circ \$_r = 0 \tag{7}$$

where "$\circ$" denotes the reciprocal product.

Assume that $\$ = [p, q, r, l, m, n]$, $\$_r = [p_r, q_r, r_r, l_r, m_r, n_r]$;

$$\$ \circ \$_r = (r - r_r) \cdot s \times s_r = -a_r \sin\alpha_r \tag{8}$$

where the two screws denote the linear vector, if $a_r = 0$ or $\alpha_r = 0$, Eq. (8)

denotes that the two screws must parallel each other or intersect to a point.

As usually, the screws are given by a commonly form. Then the reciprocal product is defined as:

$$\$ \circ \$_r = pl_r + qm_r + rn_r + p_r l + q_r m + r_r n \tag{9}$$

For a limb of a parallel manipulator, if the order of its screw system is six, there exists no reciprocal screw correspond with this limb screw system. If the order of its screw system less than six, assumed n, there exist 6-n linearly independent wrenches reciprocal with these screws, we called n order screw system.

Analyse a limb of a parallel manipulator, if there exist a n order screw system (the screws of joints are maximal linearly independent), there have 6-n reciprocal screws corresponding with it. If given the 6-n linearly independent wrenches, Eq. (9) can be used to find a n-order system of screw.

1.6.3 Identification of the Special Displacement of Limbs Structures

The screw system associated with the joint screws of a limb is called a limb twist system. The joint screws of a limb with n ($n<6$) DoF joints form a n-system. The reciprocal screw form a (6-n) system of wrenches called a limb wrench system. We called the union of the limb wrench system a platform wrench system. The platform wrench system. The wrenches of the platform wrench system restrain the movement of the moving platform. Hence, we can analyse the wrench system of the moving platform to character the movement of moving platform.

In this paper, we limit to the special displacement of those 3-DoF parallel manipulators based on literatures[11~13]. In order to supplement the structures (it assumed that l is not equal to zero in literatures) which analyzed by Fang, we conclude the special displacement with two cases (l is equal to zero in two cases).

Case one The moving platform restrained by1-, 2-, 3-pure linearly independently couple respective, the movement of the moving platform is translational.

Case two The moving platform restrained by1-, 2-, 3-pure linearly independently force respective, the movement of the moving platform is rotational.

1.6.3.1 Analysis of the structure of limbs with Infinite Pitch Wrench (IPW)

As discussed previously, if a reciprocal screw with the IPW, it denote as a pure couple which will restrict the rotational movement of platform. If a reciprocal screw satisfy with $\boldsymbol{\$}^{\mathrm{T}} \cdot \boldsymbol{\$}_0 = 0$, it denote a pure force provided by limbs which restrict translation of the moving platform. We consider if a limb provide one reciprocal screw restraint to the moving platform, the screw system of this limb have five screws which linearly independent each other. Based on the reciprocal theory, a reciprocal screw can be obtained, these wrenches will determine the movement property of the moving platform.

Constraint couple lies in *YZ* plane

Let the unit wrench of infinite pitch assumed by the following special form:

$$\hat{\boldsymbol{\$}}_r = \begin{bmatrix} 0 \\ s \end{bmatrix} = [0,0,0,0,m_r,n_r]^{\mathrm{T}} \tag{10}$$

where $m_r^2 + n_r^2 = 1$. With one restrain, assume $m_r \neq 0$, all feasible screws of a limb form a five system. Solving Eq. (7) for the reciprocal screws, we obtain five basis twists:

$$\boldsymbol{\$}_1 = \left[0, -\frac{n_r}{m_r}, 1, 0, 0, 0\right]^{\mathrm{T}},\ \boldsymbol{\$}_2 = [1,0,0,0,0,0]^{\mathrm{T}},$$
$$\boldsymbol{\$}_3 = [0,0,0,1,0,0]^{\mathrm{T}},\ \boldsymbol{\$}_4 = [0,0,0,0,1,0]^{\mathrm{T}},$$
$$\boldsymbol{\$}_5 = [0,0,0,0,0,1]^{\mathrm{T}}$$

A general screw in the five-system can be written as a linear combination of the above basis twists, namely:

$$\boldsymbol{\$} = a\ \boldsymbol{\$}_1 + b\ \boldsymbol{\$}_2 + c\ \boldsymbol{\$}_3 + d\ \boldsymbol{\$}_4 + e\ \boldsymbol{\$}_5 = \left[b, -\frac{an_r}{m_r}, a, c, d, e\right] \tag{11}$$

where a、b、c、d、e are arbitrary constants, which can not be simultaneously equal to zero. There are two special cases.

Case one Letting $a = b = 0$ and normalize the screw, Eq. (11) reduces to a unit twist of infinite pitch.

$$\hat{\$}_r = \begin{bmatrix} 0 \\ s \end{bmatrix} = \frac{1}{\omega}[0,0,0,c,d,e]^{\mathrm{T}} \tag{12}$$

where $\omega = \sqrt{c^2 + d^2 + e^2}$, since c、d、e are arbitrary constants, the prismatic joint axis can be oriented arbitrarily as long as they are linearly independent.

Case two Applying the conditions $s^{\mathrm{T}} \cdot s_0 = 0$ and $s^{\mathrm{T}} \cdot s = 1$, Eq. (11) reduces to a unit twist of zero pitch.

$$\hat{\$}_r = \begin{bmatrix} 0 \\ s \end{bmatrix} = \frac{1}{\omega}\left[b, -\frac{an_r}{m_r}, a, c, d, \frac{dn_r}{m_r} - \frac{bc}{a}\right]^{\mathrm{T}} \tag{13}$$

where $\omega = \sqrt{b^2 + \frac{a^2 n_r^2}{m_r^2} + a^2}$.

We also observe from Eq. (10) and Eq. (12) that $s^{\mathrm{T}} \cdot s_r = 0$.

Conclusion: As the constraint with couple lies in the *YZ* plane, the characterize of movement of the moving platform is a type of movement with 4-DoF (3 translational and 1 rotational motion), and this kind of movement must satisfy the condition point out as following.

Case one. All the revolute joints of a C-limb are perpendicular to the given wrench axis.

Case two. The prismatic joint axes, if any, of a C-limb can be oriented arbitrarily as long as linearly independent.

In geometry, the structure of joints must satisfy conditions given by following.

$$x = \frac{bz - d}{a}, \quad y = \frac{1}{a}\left(c - z\frac{an_r}{m_r}\right) \tag{14}$$

where z is an arbitrary parameter, (x, y, z) is a vector from the original point of Cartesin coordinate system to the axis of joints. a、b、c、d are all arbitrary parameters, n_r、m_r is a certain parameter given by the constraint system.

Constraint couple lies on *z* axis

The wrench in this case assumed the following special form:

$$\hat{\$}_r = \begin{bmatrix} 0 \\ s \end{bmatrix} = [0,0,0,0,0,n_r]^{\mathrm{T}} \tag{15}$$

where n_r is constant number, platform with one constrain which direction along the z axis, all feasible screws of a limb form a five order screw system. Solving the Eq. (7), above describe for the reciprocal screws, we obtain five basis screws:

$$\$_1 = [0,1,0,0,0,0]^{\mathrm{T}},\ \$_2 = [1,0,0,0,0,0]^{\mathrm{T}},$$
$$\$_3 = [0,0,0,1,0,0]^{\mathrm{T}},\ \$_4 = [0,0,0,0,1,0]^{\mathrm{T}},$$
$$\$_5 = [0,0,0,0,0,1]^{\mathrm{T}}$$

A general screw in this five order system can be written as a linear combination of the above basis screws, namely.

$$\$ = a\ \$_1 + b\ \$_2 + c\ \$_3 + d\ \$_4 + e\ \$_5 = [b,a,0,c,d,e] \tag{16}$$

where a、b、c、d、e are arbitrary constants, which can not be simultaneously equal to zero. There are two special cases.

Case one Letting $a = b = 0$ and predigest the vector to an unit, Eq. (16) of above reduces to an unit twist of infinite pitch.

$$\hat{\$}_r = \begin{bmatrix} 0 \\ s \end{bmatrix} = \frac{1}{\omega}[0,0,0,c,d,e]^{\mathrm{T}} \tag{17}$$

where $\omega = \sqrt{c^2 + d^2 + e^2}$, since c、d and e are arbitrary constants, the prismatic joint axis can be oriented arbitrarily as long as they are linearly independent.

Case two Applying the conditions $s^{\mathrm{T}} \cdot s_0 = 0$ and $s^{\mathrm{T}} \cdot s = 1$, Eq. (16) reduces to an unit twist of zero pitch.

$$\hat{\$}_r = \begin{bmatrix} s \\ s_0 \end{bmatrix} = \frac{1}{\omega}[b,a,0,c,-\frac{bc}{a},e]^{\mathrm{T}} \tag{18}$$

where $\omega = \sqrt{b^2 + a^2}$.

We also observe from Eq. (10) and Eq. (13) that $s^{\mathrm{T}} \cdot s_r = 0$.

Conclusion: Differ from the result of most researcher analyzed, the constraint couple axis points along z axis denote two rules following.

Rule one If the limb composed by revolute joints and all of axis of joints must parallel with the plane XY, however, when the Eq. (18) taken into accout, this instance will be inexistence.

Rule two The axis of prismatic joints must intersect a point lie on the z axis, in this situation, based on geometry analysis with joints, this instance will also inexistence.

From two rules previous described, we can conclude that there must not exist five redundant DoF in 3-DoF parallel manipulators.

1.6.3.2 Analysis of the structure of limbs with Finite Pitch Wrench (FPW)

We have discussed the structure of limbs when it constrained by a pure couple. At this paragragh we will discuss the structure of limbs when a pure force act upon the platform. Described as previous, the pure force provides a force constraint to platform and its general form can denoted as following.

$$\hat{\$}_{r1} = \begin{bmatrix} \boldsymbol{s} \\ \boldsymbol{r} \times \boldsymbol{s} \end{bmatrix} = [0, m_r, n_r, y_r n_r - z_r m_r, -x_r n_r, x_r m_r] \tag{19}$$

where direction of the constraint force lie in the plane of YZ. (x, y, z) is a vector point from the original point to any point on the wrench.

Another form about the force constraint acting on the platform can denoted as following.

$$\hat{\$}_{r2} = \begin{bmatrix} \boldsymbol{s} \\ \boldsymbol{r} \times \boldsymbol{s} \end{bmatrix} = [0, 0, n_r, y_r n_r, -x_r n_r, 0] \tag{20}$$

where the direction of the constraint force lie on the aixs of z.

Constraint the platform with pure force which axis lie on the plane of *YZ* and *z* axis

As the axis of pure force lie on the plane of YZ, so it can constraint the transfer of YZ plane. Then there have another problem bring out, the constraint force along the certainly axis lie in the plane YZ, and it will constrain the platform incompletely constrain of rotation. This incompleteness constrain denoted as following.

The force constraint the transfer along $\hat{\$}_{r1}$ direction, that is to say the force also demand velocity superposition with (s_r, s_0) along this direction of any rigid which movement is rotation is equal to zero. So the platform will rotation around the axis superposition with the axis of force or parallel it.

Case one The reciprocal of the Eq. (20) can denote by five basis screws, any screw will be assembled by these basis screws, the basis screws denote as following.

$$\$_1 = [0,1,0,0,-x_r,0]^{\mathrm{T}},\ \$_2 = \left[0,1,0,0,\frac{n_r x_r}{m_r},0\right]^{\mathrm{T}},$$

$$\$_3 = \left[0,0,1,0,-\frac{n_r y_r - m_r z_r}{m_r},0\right]^{\mathrm{T}},\ \$_4 = \left[0,0,0,0,-\frac{n_r}{m_r},1\right]^{\mathrm{T}},$$

$$\$_5 = [0,0,0,0,0,1]^{\mathrm{T}}$$

A general screw in this five order system can be written as a linear combination of the above basis screws, namely:

$$\begin{aligned} \$ &= a\ \$_1 + b\ \$_2 + c\ \$_3 + d\ \$_4 + e\ \$_5 \\ &= \left[a,b,c,0,\frac{-am_r x_r + bn_r x_r - cn_r y_r + cm_r z_r - dn_r}{m_r},d+e\right] \end{aligned} \tag{21}$$

where a、b、c、d、e are arbitrary constants, which can not be simultaneously equal to zero. There are two special rules.

Rule one Letting $a = b = c = 0$, and predigest the vector to an unit, then the Eq. (21) reduce to an unit screw of infinite pitch.

$$\hat{\$}_r = \begin{bmatrix} 0 \\ s \end{bmatrix} = \frac{1}{\omega}\left[0,0,0,0,-\frac{dn_r}{m_r},d+e\right]^{\mathrm{T}} \tag{22}$$

where $\omega = \sqrt{\frac{d^2 n_r^2}{m_r^2} + (d+e)^2}$, we also deduce that the prismatic joint axis must be oriented the direction of s.

Rule two When $d + e = \frac{-b(-am_r x_r + bn_r x_r - cm_r y_r + cm_r z_r - dn_r)}{cm_r}$, the Eq. (22) satisfy the $s^{\mathrm{T}} \cdot s_0 = 0$ and it denote a linear vector with its pitch equal to zero.

Case two The reciprocal of the Eq. (20) can denote by five basis screws, any screw will assembled by the basis screws, the basis screws denote as following.

$$\$_1=[0,1,0,0,0,0]^{\mathrm{T}},\ \$_2=[0,1,0,0,0,x_r]^{\mathrm{T}},$$
$$\$_3=[1,0,0,0,0,-y_r]^{\mathrm{T}},\ \$_4=[0,0,0,0,1,0]^{\mathrm{T}},$$
$$\$_5=[0,0,0,1,0,0]^{\mathrm{T}}$$

There also have a general screw by a linear combination of these five basis screws, namely:

$$\$=a\ \$_1+b\ \$_2+c\ \$_3+d\ \$_4+e\ \$_5=[a,b,c,d,e,bx_r-cy_r] \quad (23)$$

where a、b、c、d、e are arbitrary constants, which can not be simultaneously equal to zero. There are two special rules.

Rule one Letting $a=b=c=0$, and predigest the vector to an unit, then the Eq. (23) reduce to an unit screw of infinite pitch $\$=\begin{bmatrix}0\\s\end{bmatrix}=\frac{1}{\omega}[0,0,0,f,e,0]$.

where $\omega=\sqrt{f^2+e^2}$, we also deduce that the prismatic joint axis must be oriented the direction of $\boldsymbol{s}$.

Rule two When $e=\dfrac{c(cy_r-bx_r)-af}{b}$, the Eq. (23) satisfy the $\boldsymbol{s}^{\mathrm{T}}\cdot\boldsymbol{s}_0=0$ and it denote a linear vector which its pitch equal to zero.

Analysis of the structure of joints on the lamb (with pure force and *YZ* plane)

As above, we have gained the type of screws basis system which reciprocal with the original wrench. Then we consider the different situation with the rotational and translation pairs.

When the reciprocal wrench lie in *YZ* plane, the axis of rotational and translation pairs will be denoted by two rules.

Rule one When the pitch which combinated by basis screws is infinite, its axis will lie in the plane *YZ*, it can not perpendicular to the plane *YZ*. So the only displacement of the axis of the prismatic is satisfying the equation as following: $\frac{d+e}{d} = -\frac{n_r^2}{m_r^2}$.

Rule two When the pitch of the combination of basis screws is zero, its axis intersect a point which displacement depend on the vector as following: $x = \frac{az - N}{c}$, $y = \frac{bz}{c}$.

where $N = \frac{-am_r x_r + bn_r x_r - vn_r y_r + cm_r - dn_r}{m_r}$, z is arbitrary number.

Analysis of the structure of joints on the limb (with pure force and the axis of z)

We will consider the different displacement of the axis of rotational and translation pairs. There classify two different situation as following:

Rule one When the pitch combinated by basis screws is infinite, the form of combination screws lie in the plane of XY. Its always perpendicular with the axis of z. So in this situation, it must satisfy f and e simulation equal to zero.

Rule two When the pitch combinated by basis screws is zero, its axis intersect a point which displacement depend on the vector as following: $x_r = y_r$, $x = \frac{a}{b}y$, $z = \frac{cy - f}{b}$.

1.6.4 Conclusions

In this paper, the theory of reciprocal screws was used to develop a systematic approach for structure synthesis of 3-DoF parallel manipulators with revolute and prismatic joints respectively. It has shown that when the axis of reciprocal screw located in the plane yz under the Cartesian Coordinate system, the motion property of moving platform are derived. Conclusion based on different displacement of the axis of joints is that the degree of freedom of platform will change. Instantaneous and non-instantaneous DoF involved to analysis the characterize of the movement of platform. There also discussed the redundancy DoF in this paper.

References

[1] Steward D. A Platform with Six Degree of Freedom [J]. Proceeding Institute of Mechanical Engineering. 1965, Vol. 180, No. 5: 371 ~ 386.

[2] Hunt K H. Kinematic Geometry of Mechanisms [M]. London, England: Oxford University Press, 1978.

[3] R Clavel DELTA. A fast robot with parallel geometry. Proceedings of 18th International Symposium on Industrial Robot, Lausanne, Switzerland, April 1988, 91 ~ 100.

[4] Tsai L W, WALSH G C and STAMPER R E. Kinematics of a novel three DoF translational platform [C] //Proceedings of the 1996 IEEE International Conference on Robotics and Automation, Minneapolies, MN, 1996, 3446 ~ 3451.

[5] Tsai L W. Systematic Enumeration of Parallel Manipulators: in Parallel Kinematics Machines [J]. Edited by C. R. Boer, L. Molinari-Tosatti, and KS. Smith, Springer New York, 1999: 33 ~ 50.

[6] Gregorio R D. Kinematics of the Translational 3-URC Mechanism [C] //Proceedings of IEEE/ASME International Conference on Advanced Intelligent Mechatronics, Como, Italy, 2001: 147 ~ 152.

[7] Tsai L W. Kinematics of a Three-DoF Platform Manipulator with Three Extensible Limbs [J]. In Recent advances in robot Kinematics, edits by J. Lenarcic and V. Parenti-Castelli, London: Kluwer Academic Press. 1996: 401 ~ 410.

[8] Herve J M and Sparacino F. Structural synthesis of parallel robots generating spatial translation [C] //IEEE International Conference on Robotics and Automation, Italy: Pisa, June 1992, 808 ~ 813.

[9] Frisoli A D, Checcaci D, Salsedo E and Bergamasco M. Synthesis by screws algebra of translating in parallel actuated mechanisms [J]. Advances in robot kinematics, edited by J. Lenarcic and M. M. Stanisic, Boston: Kluwer Academics, 2000: 433 ~ 440.

[10] Kong X and Gosselin C. Generation of parallel manipulators with three translational degrees of freedom based on screw theory [C] //Proceedings of IFToMM Symposium on Mechanisms, Machines, and Mechatronics, Canada: Saint Hubert. 2001.

[11] Fang Y F, Tsai L W. Analytical identification of limb structures for translation parallel manipulators [J]. Journal of Robotic Systems, 2004, 21 (5): 209 ~ 218.

[12] Fang Y F, Tsai L W. Structure synthesis of a class of 3-DoF rotational parallel manipulators [J]. IEEE Translations on Robotics and Automation, 2004, 20 (1): 117 ~ 121.

[13] Fang Y F, Tsai L W. Structure synthesis of a class of 4-DoF and 5-DoF parallel manipulators with identical limb structures [J]. International Journal of Robotics Research, 2002, 21 (9): 799 ~ 810.

2 并联机构智能控制系统研究

2.1 基于模糊神经网络运算法则的并联机器人自适应控制研究

本文针对并联机器人数学模型不完全确知并包含外部扰动的非线性多变量系统，提出一种基于模糊神经网络运算法则（FNNA）的自适应控制策略。将各个支链的模糊规则通过神经网络进行在线训练并得出模糊规则的权重并将此运用于在线辨识非线性自适应控制系统的未知动态，有效抑制了系统的数学模型不精确所产生的误差及外部扰动。仿真结果表明该控制方法明显提高了控制系统的轨迹跟踪性能，并对外部干扰及系统的非线性具有很强的鲁棒性。

2.1.1 引言

在并联机器人控制系统中，由于存在模型的不确定性、系统的非线性以及外界的干扰等因素，所以，在此类并联机构的控制系统设计方面，必须要考虑其鲁棒特性及其控制系统的稳定性。为了消除机构的不确定性和外界扰动对控制系统的影响，Takegaki 和 Arimoto[1] 于 1981 年提出一种基于反馈方式对机器人动态模型进行控制系统设计。Tomei[2] 采用自适应 PD 控制策略进行控制系统设计。在以上的控制系统设计中，机器人不确定性参数都被设定为一定范围的定值，这些假设与机器人实际运行的参数不完全相符合。作为对以上控制系统设计方法的完善，一系列基于线性自适应控制策略被提出[3]。然而，由于机器人并联机构是一个典型的非线性系统以及由于在线计算量大等因素，这些控制方法无法获得预期的控制品质。为了解决这些问题，Vicente 等[4] 提出基于连续滑模 PID 控制系统的设计，并在该控制系统的稳定性和收敛性方面做了较深入的研究。为了消除并联机构中参数的不确定性以及外界干扰对控制系统的影响，Cheng 等[5] 提出采用自适应变结构控制方法对并联机构的控制系统进行设计，并基于Lyapounov稳定性理论解决了该控制器轨迹跟踪鲁棒性问题。Byung和Woon[6]提出一种采用模糊补偿器的自适应控制系统的设计方法，从而解决了由于摩擦、模型的不精确以及外界的干扰等对控制器的影响。本文提出一种基于模糊神经网络算法的自适应控制系统的设计方法。该控制系统分为两个主要部分：第一部分是采用自适应滑模控制方法，以期对并联机构的多输入多输出(MIMO)系统进行鲁棒控制；第二部分采用了模糊神经网络算法来解决误差的在线修正、系统的稳定性以及轨迹跟踪。在建立液压驱动器的数学模型的基础上，本文分别对控制系统的设计以及控制系统的稳定性进行分析，并给出了仿真实验结果。仿真实验结果表明，采用该控制方法可明显提高控制系统的跟踪性能，并具有很强的对外界干扰抑制能力，具有较强的鲁棒性。

2.1.2 并联机构液压伺服驱动器数学模型

在不考虑弹性载荷的情况下，连续的非对称电位伺服液压缸数学模型可表示为：

$$Y(s)=\frac{\dfrac{K_{x\alpha}}{A_{me}}x-\dfrac{K_{t\alpha}}{A_eA_{me}}\left(\dfrac{V_e}{4\beta_eK_{t\alpha}}s+1\right)f_e}{\left(\dfrac{s^2}{\omega_{he}^2}+\dfrac{2\xi_{he}}{\omega_{he}}+1\right)s} \tag{1}$$

式中，$K_{x\alpha}$为流动增益，m^2/s；A_{me}为活塞平均面积，m^2；$K_{t\alpha}$为总压力系数，$m^2/(s\cdot N)$；A_e为液压缸当量面积，m^2；β_e为弹性模数，MPa；f_e为等效外界干扰力，N；x为阀体尺寸，m。

$$\omega_{he}=2\sqrt{\frac{A_eA_{me}\beta_e}{V_e\mathrm{m}}}\quad \mathrm{rad/s} \tag{2}$$

$$\xi_{he}=K_{t\alpha}\sqrt{\frac{m\beta_e}{V_em}}+\frac{B_p}{4}\sqrt{\frac{V_e}{m\beta_eA_eA_{me}}}$$

式中，m为负载质量，kg；B_p为负载黏性阻尼系数，N·s。

液压伺服系统模型如图1所示。

由上可知，具有干扰力作用的非对称液压缸的数学模型与对称液压缸的相类似。当活塞沿液压杆方向运动且阀门不在零点位置时，非

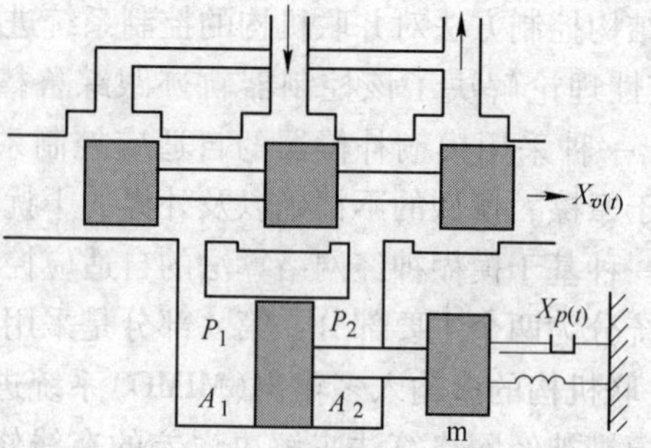

图 1 液压伺服系统模型

对称液压缸的运动等效于带有附加力的对称液压缸的负载形式，可表示为非对称液压缸无杆腔内的压力 P_1 乘以液压杆面积 α。同理，当活塞在无杆腔内运动时，相应的外部等效干扰力可以通过相同的方法得到。因此，事实上系统的输出量由两方面的输入所决定：一是阀门的位置 x；二是等效的外部干扰力 f_e。

液压伺服驱动系统传递函数框图如图 2 所示。

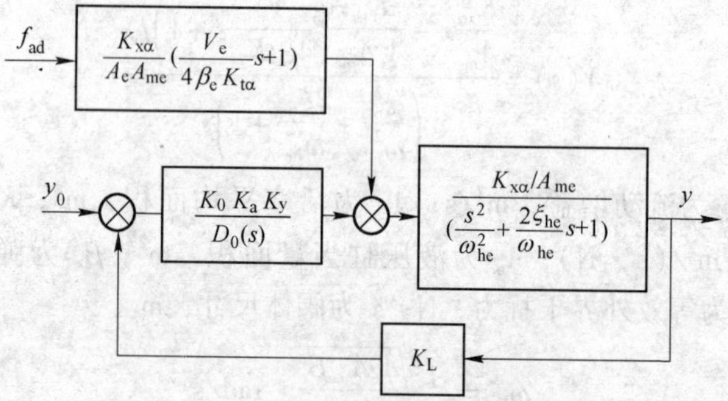

图 2 液压伺服驱动系统传递函数框图

图 2 中，$f_{ad}=\begin{cases}\dfrac{\eta^2 ap_s}{1+\eta^2}, & x>0\\[2ex] \dfrac{ap_s}{1+\eta^2}, & x<0\end{cases}$；$\eta=\dfrac{A_2}{A_1}$。

并联机器人原理图如图 3 所示。

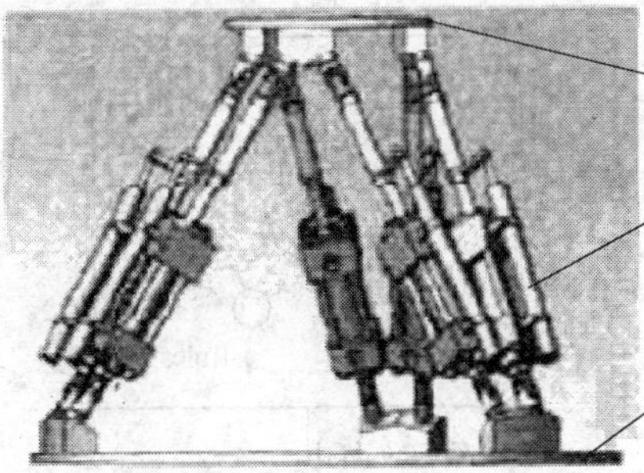

图3 并联机器人原理图

2.1.3 具有模糊神经网络运算法则的自适应控制器设计

2.1.3.1 自适应控制器结构

在许多场合下，控制对象的数学模型未知或难以用确切的数学公式表达出来，对于具有多输入多输出的控制系统则更显得尤为突出。为了解决这方面的问题，本文提出在自适应控制系统中采用模糊神经网络运算法则，以解决并联机器人多输入多输出这种具有典型的非线性耦合所带来的模型不确定情况下控制系统设计问题，并讨论了所设计的控制系统的稳定性问题。具有模糊神经网络运算法则的自适应控制系统框图如图 4 所示。

本文所提出的自适应控制器具有自适应控制器和模糊神经网络运算法则两个部分。为了降低实际输出与理想输出之间的误差，本文采用了基于李亚普诺夫稳定性理论基础上的自适应鲁棒控制律。由于并联机器人的模型为典型的多输入多输出非线性系统，模糊运算法则需要大量的模糊规则基，从而使得系统在模糊推理方面显得复杂，模糊规则基在系统中所占的权重则影响了系统的精确度（如果对于 n 自由度并联机器人而言，输入变量选择 k，则需要 nk^{3n} 模糊规则基）。因此，采用神经网络技术，通过在线修正各模糊规则基在控制系统中的权值以提高控制系统的精确度。

2.1.3.2 模糊神经网络运算法则

模糊逻辑系统通常是由模糊规则基、模糊推理、模糊化算子和非

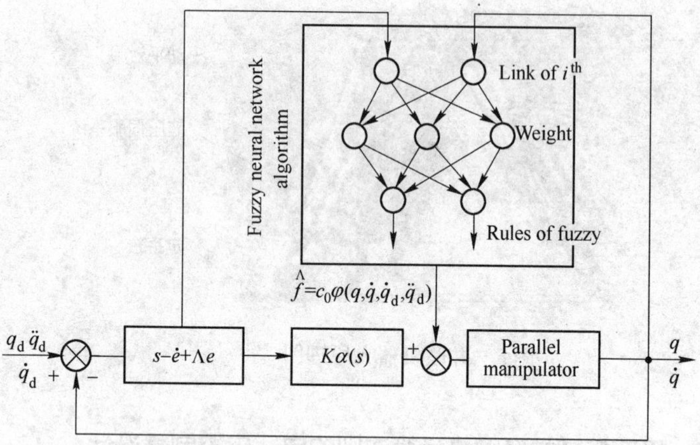

图 4　具有模糊神经网络运算法则的自适应控制系统框图

模糊化（解模糊化算子）四部分所组成。本文所研究的多输入多输出系统的模糊规则基定义为如下形式：

$$R = U_{l=1}^{M} R_l \tag{3}$$

每个 R_l 规则表示为：

$$R_l: \text{if}\ \ x_1 \text{ is } A_1^l\ \ x_2 \text{ is } A_2^l\ \ \cdots\ \ x_n \text{ is } A_n^l, \text{then}\ \ y_1 \text{ is } B_1^l\ \ y_2 \text{ is } B_2^l\ \ \cdots\ \ y_m \text{ is } B_m^l \tag{4}$$

式中，M 表示模糊规则的总数；$x = (x_1, x_2, \cdots, x_n)^T$ 和 $y = (y_1, y_2, \cdots, y_m)^T$ 分别表示模糊系统的输入与输出矢量；A_i^l 和 B_i^l 分别为模糊器中子系统 U_i 和 V_j 的模糊语言描述；其中：$U = U_1 \times U_2 \times \cdots \times U_n$，$U_i \in \mathbf{R}$；$V = V_1 \times V_2 \times \cdots \times V_m$，$V_j \in \mathbf{R}$，$i = 1, 2, \cdots, n$，$j = 1, 2, \cdots, m$。

用阈值函数 $\mu_{A_i^l}(x_i)$ 以及 $\mu_{B_j^l}(x_j)$ 来表示各个模糊规则基所产生的作用。单个模糊器的形式表示为：

$$\boldsymbol{y}_j = \frac{\sum_{l=1}^{M}\left[\prod_{i=1}^{n}\mu_{A_i^l}(x_i)\right]\boldsymbol{y}_j^{-1}}{\sum_{l=1}^{M}\left[\prod_{i=1}^{n}\mu_{A_i^l}(x_i)\right]} \tag{5}$$

由于并联机器人的非线性及强耦合特性，所以在机构模型参数中存在大量的未知参数及不确定因素。因此，本文在模糊控制中融合了神经网络运算法则，通过在线调整各个模糊基的权重以达到控制的最优，使设计出来的控制系统具有强的鲁棒性。

设各个模糊基所对应的权值为 ε_i^M，则式（5）可改写为如下形式：

$$\boldsymbol{y}_j = \sum_{j=1}^{M} \boldsymbol{y}_j^{-1} \boldsymbol{\varepsilon}(x) = \theta_j^{\mathrm{T}} \boldsymbol{\varepsilon}(x) \tag{6}$$

式中，$\boldsymbol{\varepsilon}(x) = \dfrac{\prod_{i=1}^{n} \mu_{A_i^l}(x_i)}{\sum_{l=1}^{M} \left[\prod_{i=1}^{n} \mu_{A_i^l}(x_i) \right]}$

$\boldsymbol{\varepsilon}(x) = (\varepsilon_1(x), \varepsilon_2(x), \cdots, \varepsilon_M(x))^{\mathrm{T}} \in \mathbf{R}^M$ 称为模糊基函数矢量；$\boldsymbol{\theta}_j = (y_j^{-1}, y_j^{-2}, \cdots, y_j^{-M})^{\mathrm{T}} \in \mathbf{R}^M$ 称为参数矢量；在神经网络隐含层中固定映射为 $\boldsymbol{\varepsilon}(x)$，则可调整权值 $\boldsymbol{\theta}_j$。

2.1.4 仿真研究

采用6-RPS 型并联机器人作为仿真研究对象，该并联机器人具有三个自由度并通过移动作为驱动器。驱动器的数学模型由 2.1.2 节所述方法建立。仿真实验数据确定如下：

液压伺服驱动器参数为：

$$\alpha_2 \in [100, 170], \alpha_1 \in [90000, 100000], \alpha_0 \in [5800, 6000],$$
$$\beta \in [5800, 6000], F \in [-1000, 1000]$$

设计期望跟踪的轨迹为两转动和一移动轨迹，期望跟踪轨迹函数如下：

$$\varphi_{1d} = (2.5\pi/12)\sin t, d = -800t^4 + 1100t^3 - 200t^2, \varphi_{2d} = (3.75\pi/12)\cos t$$

具有模糊神经网络运算法则的控制器参数设定为：

$$K_0 = 85,\ K_1 = 0.8,\ K_2 = 0.005$$

阶跃信号作为并联机器人轨迹输入信号。

仿真结果如图 5 ~ 图 10 所示。

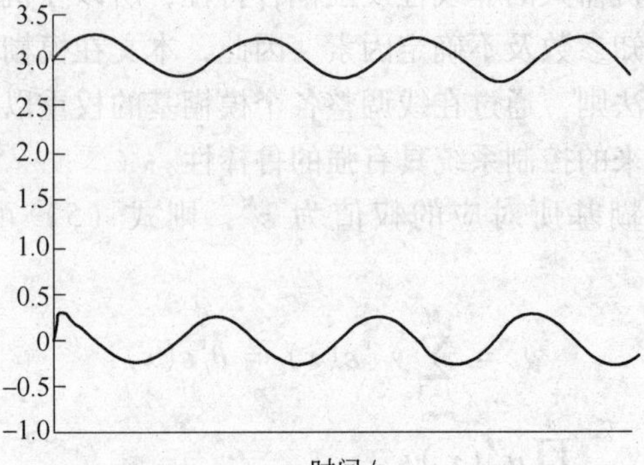

图 5 三液压伺服驱动器轨迹跟踪

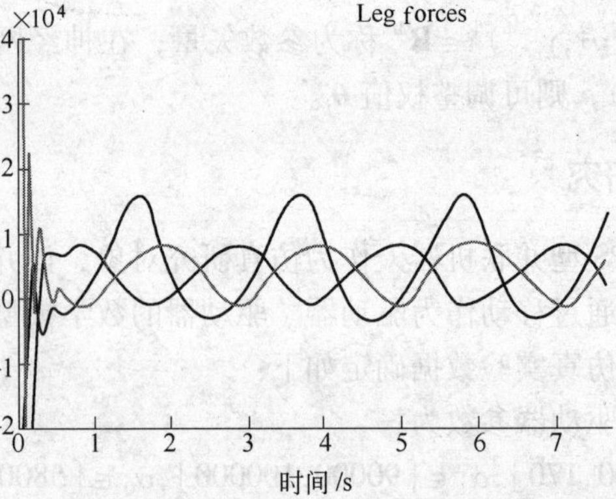

图 6 6-RPS 型并联机器人支链受力轨迹跟踪

2.1.5 结论

本文针对 6-RPS 型三自由度并联机器人提出了具有模糊神经网络运算法则的自适应控制器的设计方案。考虑所控制对象的非线性特性以及强耦合性，运用神经网络在线调整该多输入多输出的模糊基在控制系统中的权值，并通过对并联机器人数学模型系统仿真说明所设计的模糊自适应控制器具有较高的鲁棒性。

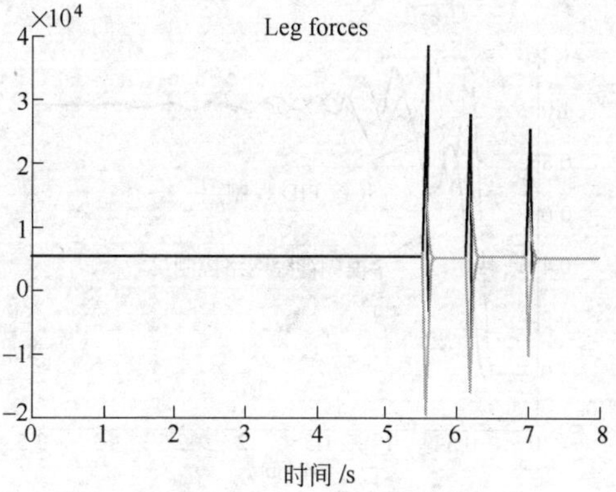

图7　具有扰动的支链受力轨迹跟踪

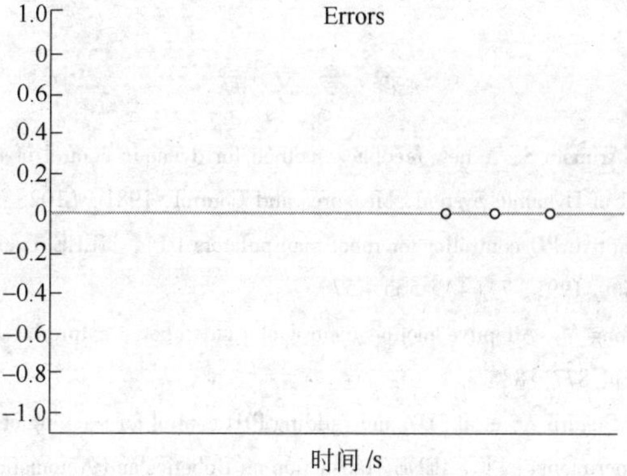

图8　具有扰动的误差轨迹跟踪

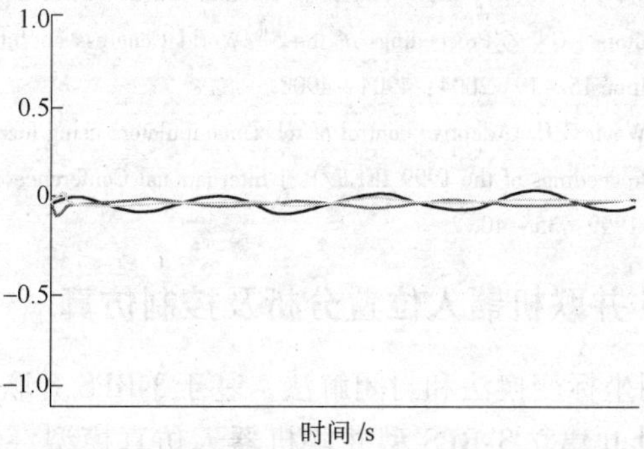

图9　三液压伺服驱动器误差轨迹跟踪

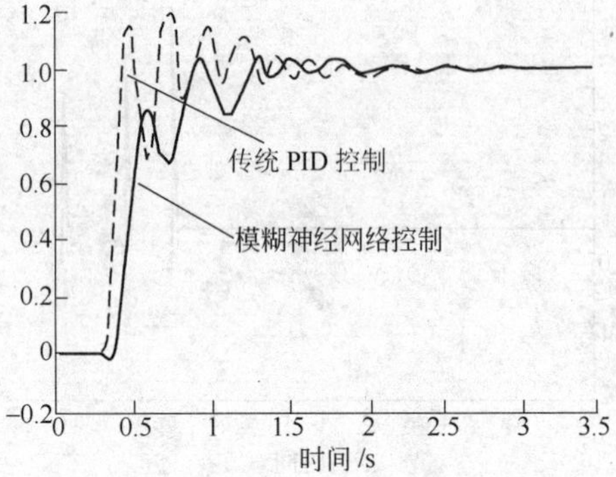

图 10 模糊神经网络控制与传统 PID 控制轨迹跟踪比较

参考文献

[1] Takegaki M, Arimoto S. A new feedback method for dynamic control of manipulators [J]. ASME Journal of Dynamic System, Measures and Control, 1981, (102): 119 ~ 125.

[2] Tomei P. Adaptive PD controller for robot manipulators [J]. IEEE Transaction on Robotica and Automation, 1991, 7 (4): 565 ~ 570.

[3] Ortega R, Spong M. Adaptive motion control of rigid robots: a turorial [J]. Automatica, 1989, 25 (6): 877 ~ 888.

[4] Vicente P V, Sugurn A, et al. Dynamic sliding PID control for tracking of robot manipulators: theory and experiments [J]. IEEE Transaction on Robotics and Automation, 2003, 19 (6): 967 ~ 976.

[5] Cheng C, Chien S H, et al. Design of adaptive variable structure controllers with application to robot manipulators [C] //Proceedings of the 5th World Congress on Intelligent Control and Automation, June 15 ~ 19, 2004: 4904 ~ 4908.

[6] Byung K Y, Woon C H. Adaptive control of robot manipulators using fuzzy compensator, Part one [C] //Proceedings of the 1999 IEEE/RSJ International Conference on Intelligent Robots and Systems, 1999: 35 ~ 40.

2.2 3-RPS 并联机器人位置分析及控制仿真

本文应用坐标变换法和封闭解法，基于 3-RPS 并联机器人位置分析，采用 Matlab 建立 3-RPS 型并联机器人仿真模型，给定不同状态下的参考输入，通过合理设置 PID 控制器各项参数进行运动学仿真比较。仿真结构表明：在不同的参考输入条件下，动平台的位置变化几乎相同，实现了PID控制调平作用，驱动杆实际输出与参考值的偏差在允许范围内波动，控制效果明显，为研究基于并联机器人自动调平提供参考。

2.2.1 引言

并联机构的位置分析在并联机器人的研究中具有十分重要的意义，通过位置分析可以求解机构的输入与输出构件之间的位置关系，是机构运动分析最基本的任务，也是机构受力分析，速度、加速度分析，误差分析，机构综合分析等的基础[1]。由于并联机器人机构具有刚度高、承载能力强、结构紧凑等优点，所以它在工业生活中具有广阔的应用前景。少自由度并联机器人机构运动学、动力学研究相对简单，具有很强的灵活性，制造相对容易，有诸多优越性，在各个领域中更具有应用潜力，因而成为国内外学者研究较多的一类机构。在串联机器人的机构位置分析中，正解比较容易，反解比较困难，相反在并联机构位置分析中反解相对简单，而位置正解比较复杂。

目前，国内外对并联机器人机构位置正解主要有解析法和数值法，它们各有自己的优缺点。数值法数学模型相对简单，能够适用于大多数的并联机构，但它不能求出全部位置的解。国内外学者对此做了大量的研究，例如Innocentit提出了位置正解的一维搜索法，Dagupta提出了预测校正法，该方法采用3维搜索法从纯几何角度求解位置正解，有学者提出了一种逐次逼近法。解析法，即利用各种数学消元法逐一消去未知数，最终求解一个一元多项式。解析法虽能求出全部位置的解，但求解过程比较繁琐。国内对位置正解解析法的研究主要是北京邮电大学机械学研究所进行的，至今，他们分别获得了3-TPS、3-6SPSS、5-4型、6-4型、6-4台体机构、6-5型机构、5-5型机构、6-5型机构的位置正解，并验证了机构解的数目[2]。

在控制领域中，控制方法很多，有PID控制、滑模变结构控制、自适应控制、模糊控制、神经元控制等，其中PID控制是最早发展起来的控制策略之一，由于其具有算法简单、鲁棒性好、可靠性高等特点，因而在控制领域中得到广泛的应用[3]。

本文在借鉴前人的基础上，以3-RPS并联机器人为例，应用坐标变换法和封闭方法对并联机器人机构位置做了初步的分析，并运用Matlab建模与PID控制进行仿真。

2.2.2 机构描述

3-RPS并联机构的上下平台以3个分支相联，每个分支由3个运动副连接而成，固定平台的每个顶点各自连接一个转动副，转动副通过一个移动副连接一个球面副，每个球面副分别连接在移动平台的各个顶点，上下平台都是由2个正三角形组成，如图1所示。

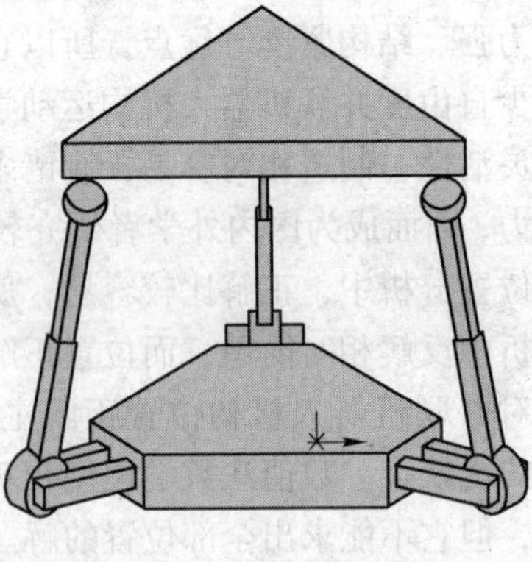

图1 3-RPS机构模型

在机构的上下平台上各建立坐标系，动坐标系 $P\text{-}UVW$ 建立在上平台，下平台建立定坐标系 $O-XYZ$，P、O 分别在上下平台的中心。上下平台分别记为 $B_1B_2B_3$，$A_1A_2A_3$，其中，PB_i、OA_i、OB_i 分别记为 b_i、a_i、q_i；b_i、a_i 的长度分别记为 h、g；d_i 表示驱动杆 A_iB_i 之间的杆长；φ_i 为连杆 A_iB_i 与下平台之间的倾斜角；L 为上平台三角形的边长，如图2所示。

2.2.3 位置分析

2.2.3.1 位置反解

位置反解，即给定上平台在空间的位置和姿态，求各个杆长，即

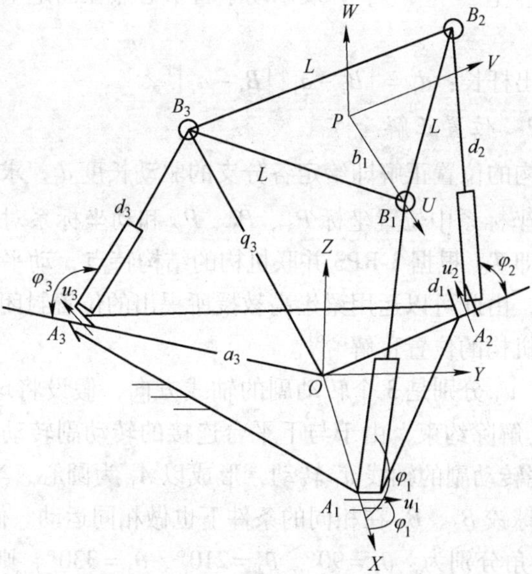

图 2　3-RPS 机构坐标示意图

各移动副的位移[4]。由图 2 可知，A_i 在 $O-XYZ$ 中的坐标为：

$$\boldsymbol{a}_1=[g,0,0]^{\mathrm{T}},\boldsymbol{a}_2=[-\frac{1}{2}g,\frac{\sqrt{3}}{2}g,0]^{\mathrm{T}},\boldsymbol{a}_3=[-\frac{1}{2}g,-\frac{\sqrt{3}}{2}g,0]^{\mathrm{T}} \quad (1)$$

B_i 在 P-UVW 中的坐标为：

$$\boldsymbol{b}_1=[h,0,0]^{\mathrm{T}},\boldsymbol{b}_2=[-\frac{1}{2}h,\frac{\sqrt{3}}{2}h,0]^{\mathrm{T}},\boldsymbol{b}_3=[-\frac{1}{2}h,-\frac{\sqrt{3}}{2}h,0]^{\mathrm{T}} \quad (2)$$

取 $Z-Y-X$ 型欧拉角（α, β, γ），则动坐标系相对固定坐标系的方向余弦为：

$$\boldsymbol{R}=\begin{bmatrix} c\alpha c\beta & c\alpha s\beta s\gamma - s\alpha c\gamma & c\alpha s\beta c\gamma + s\alpha s\gamma \\ s\alpha c\beta & s\alpha s\beta s - c\alpha c\gamma & s\alpha s\beta c\gamma - c\alpha s\gamma \\ -s\beta & c\beta s\gamma & c\beta c\gamma \end{bmatrix} \quad (3)$$

式中，$c(\cdot)=\cos(\cdot)$, $s(\cdot)=\sin(\cdot)$。

球铰在固定坐标系中的坐标为：

$$B_i=Rb_i+\boldsymbol{P} \quad i=1,2,3 \quad (4)$$

式中，$\boldsymbol{P}=[P_X \quad P_Y \quad P_Z]^{\mathrm{T}}$，表示动平台中心点在固定坐标系中的位置矢量。

由上式可求出杆长：$d_i^2=|B_i-a_i||B_i-a_i|^{\mathrm{T}}$。

2.2.3.2 位置正解

并联机构的位置正解即给定各分支的驱动长度 d_i，求动平台中心点 P 在固定坐标系中位置坐标 P_X、P_Y、P_Z 和动坐标系对定坐标系的方向余弦，即 T。根据3-RPS并联机构的结构特点，动平台由一个正三角形组成，由此可以运用梁崇高教授所提出的位置封闭解的方法来求解该并联机构的位置正解[5,6]。

u_1、u_2、u_3 分别是3个转动副的轴线方向，假设将球铰 B_1 与动平台 $B_1B_2B_3$ 解除约束，由于与下平台连接的转动副转动作用，显然杆 B_1A_1 会绕转动副的轴线 u_1 转动，形成以 A_1 为圆心，半径为 d_1 的圆。同理，球铰 B_2、B_3 在相同的条件下也做相同运动。假设 u_1、u_2、u_3 与 X 轴夹角分别为：$\theta_1=90°$、$\theta_2=210°$、$\theta_3=330°$，则上平台的3个球铰的位置分别用 φ_1、φ_2、φ_3 表示为：

$$X_{B_i}=X_{A_i}+d_i\cos\varphi_i\cos(\theta_i-90°),$$

$$Y_{B_i}=Y_{A_i}-d_i\cos\varphi_i\cos(\theta_i-90°), Z_{B_i}=d_i\sin\varphi_i$$

动平台的三个顶点坐标表示上平台三角形的边长，边长为 L，则有如下关系式：

$$(X_{B_1}-X_{B_2})^2+(Y_{B_1}-Y_{B_2})^2+(Z_{B_1}-Z_{B_2})^2=L^2,$$

$$(X_{B_2}-X_{B_3})^2+(Y_{B_2}-Y_{B_3})^2+(Z_{B_2}-Z_{B_3})^2=L^2,$$

$$(X_{B_1}-X_{B_3})^2+(Y_{B_1}-Y_{B_3})^2+(Z_{B_1}-Z_{B_3})^2=L^2$$

将上述两式整理得到以 φ_i 为参数的超越方程：

$$\begin{aligned}
&A_1\cos\varphi_1+B_1\cos\varphi_2+D_1\cos\varphi_1\cos\varphi_2+E_1\sin\varphi_1\sin\varphi_2+F_1=0\\
&A_2\cos\varphi_2+B_2\cos\varphi_3+D_2\cos\varphi_2\cos\varphi_3+E_2\sin\varphi_2\sin\varphi_3+F_2=0\\
&A_3\cos\varphi_3+B_3\cos\varphi+D_3\cos\varphi_1\cos\varphi_3+E_3\sin\varphi_1\sin\varphi_3+F_3=0 \qquad (5)
\end{aligned}$$

式中，A_i、B_i、D_i、E_i、F_i 均为已知量函数。

令 $x_i=\tan\dfrac{\varphi_i}{2}$，则有 $\sin\varphi_i=\dfrac{x_i}{1+x_i^2}$，$\cos\varphi_i=\dfrac{1-x_i}{1+x_i^2}$，代入式（5）可得：

$$[(F_1-B_1+A_1-D_1)+(F_1-B_1+A_1-D_1)x_1^2]x_2^2+(2E_1x_1)x_2+$$
$$[(F_1+B_1+A_1+D_1)+(F_1+B_1-A_1-D_1)x_1^2]=0$$
$$[(F_2-B_2+A_2-D_2)+(F_2-B_2+A_2-D_2)x_2^2]x_3^2+(2E_1x_1)x_3+$$
$$[(F_2+B_2+A_2+D_2)+(F_2+B_2-A_2-D_2)x_2^2]=0$$
$$[(F_3-B_3+A_3-D_3)+(F_3-B_3+A_3-D_3)x_1^2]x_3^2+(2E_3x_1)x_3+$$
$$[(F_3+B_3+A_3+D_3)+(F_3+B_3-A_3-D_3)x_1^2]=0$$

化简得：

$$a_1x_2^2+b_1x_2+c_1=0,\ a_2x_3^2+b_2x_3+c_2=0,\ a_3x_3^2+b_3x_3+c_3=0 \tag{6}$$

式中，a_i、b_i、c_i 为各阶变量 x_2 和 x_3 的系数。对上式进行化简消去 x_3，由于 a_i、b_i、c_i 本身只含有 x_1、x_2，则可以得到只关于 x_1、x_2 的方程：

$$(a_2c_3-a_3a_2)^2-(a_2b_3-a_3b_2)(b_2c_3-b_3c_2)=0 \tag{7}$$

将上式写成只关于 x_2 的方程有：

$$k_1x_2^4+k_2x_2^3+k_3x_2^2+k_4x_2+k_5=0 \tag{8}$$

式（8）中 k_i 仅是关于 x_1 的函数，可写成如下矩阵方程：

$$\begin{bmatrix} 0 & k_1 & k_2 & k_3 & k_4 & k_5 \\ k_1 & k_2 & k_3 & k_4 & k_5 & 0 \\ a_1 & b_1 & c_1 & 0 & 0 & 0 \\ 0 & a_1 & b_1 & c_1 & 0 & 0 \\ 0 & 0 & a_1 & b_1 & c_1 & 0 \\ 0 & 0 & 0 & a_1 & b_1 & c_1 \end{bmatrix}\begin{bmatrix} x_2^5 \\ x_2^4 \\ x_2^3 \\ x_2^2 \\ x_2 \\ 1 \end{bmatrix}=0 \tag{9}$$

由于 $[x_2^5\quad x_2^4\quad x_2^3\quad x_2^2\quad x_2\quad 1]^{\mathrm{T}}\neq 0$，所以，该齐次方程有非零解的充分必要条件是系数行列式等于零，从而求出 x_1，同理求出 x_2、x_3，从而求出 φ_i，从而得到平台 3 个顶点的坐标，因为上平台是一个等边三角形，3 个球铰在三角形的顶点上，所以根据三角形的几何性质可以求出动坐标系原点 P 的坐标。

$$P_X = \frac{1}{3}\sum_{i=1}^{3} X_{B_i}, P_Y = \frac{1}{3}\sum_{i=1}^{3} Y_{B_i}, P_Z = \frac{1}{3}\sum_{i=1}^{3} Z_{B_i} \tag{10}$$

假设动坐标系对固定坐标系的3个转角分别为 θ_U、θ_V、θ_W，则根据如下关系式求出方向余弦：

$$\cos\theta_U = \frac{1}{|PB_1|}[(X_{B_1} - P_X)i + (Y_{B_1} - P_Y)j + (Z_{B_1} - P_Z)k]$$

$$\cos\theta_V = \frac{1}{|B_3B_2|}[(X_{B_2} - X_{B_3})i + (Y_{B_2} - Y_{B_3})j + (Z_{B_2} - Z_{B_3})k]$$

$$W = U \times V \tag{11}$$

式中，$|PB_1| = \sqrt{(X_{B_1} - P_X)^2 + (Y_{B_1} - P_Y)^2 + (Z_{B_1} - P_Z)^2}$，$|B_3B_2| = \sqrt{(X_{B_2} - X_{B_3})^2 + (Y_{B_2} - Y_{B_3})^2 + (Z_{B_2} - Z_{B_3})^2}$。

2.2.4 Matlab 建模

根据控制原理与3-RPS并联机器人的结构连接关系，运用Matlab中SimMechanics模块集，建立如图3～图6的仿真模型[7~10]。

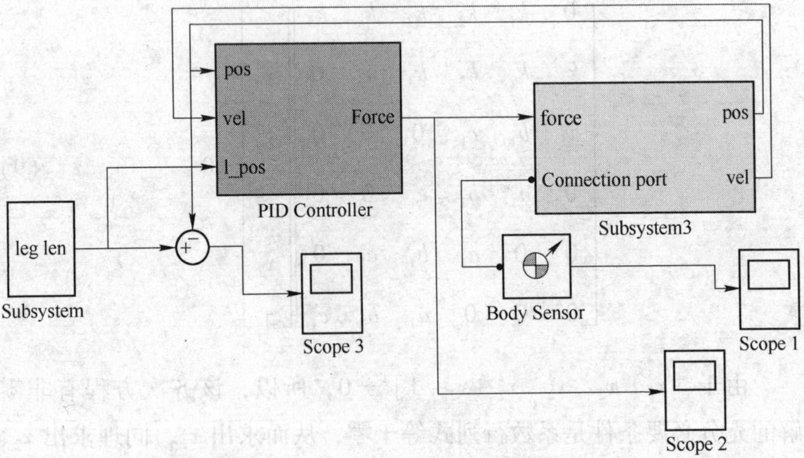

图3 3-RPS机构PID控制SimMechanics总框图

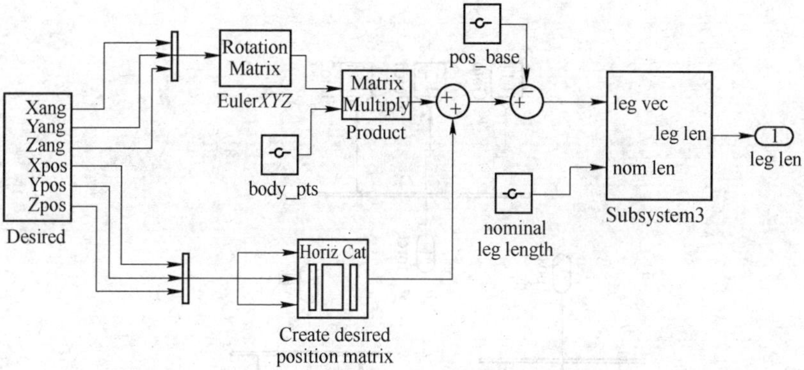

图 4 3-RPS 机构参考值输入 SimMechanics 框图

图 3 表示 3-RPS 机构建模仿真的总体框图，Subsystem 部分表示对该机构系统输入的参考值，如图 4 所示。PID Controller 部分表示控制部分；Subsystem3 部分表示该机构，如图 5 所示。Scope1 与 Scope3 为示波器，分别表示动平台的位置变化与差值的输出。

2.2.5 仿真对比分析

设定初始参数：上下平台外接圆半径分别为 50mm、100mm，上平台的质量为 1kg，移动副固定杆和移动杆都为 0.1kg，上平台中心点 P 的坐标为(0,0,150)，$\varphi_1=\varphi_2=\varphi_3=120°$，建立机构模型。

假设动平台输入的参考值：EulerXYZ(45°,45°,0°) P 点的坐标输入为：$(\sin(2\pi t),\sin(2\pi t),\sin(2\pi t))$，进行仿真，仿真结果如图 7 及图 8 所示。

改变 P 点的参考坐标输入：$(20\sin(2\pi\times5t+10)+10,20\sin(2\pi\times5t+10)+10,20\sin(2\pi\times5t+10)+10)$，进行仿真，仿真结果如图 9 及图 10 所示。

比较图 7 和图 9 可知，在不同的参考值输入的条件下，通过合理改变 PID 控制器参数，动平台中心点 P 的 X、Y、Z 轴的实际位置变化规律几乎相同，体现了 PID 控制自动调平作用。由图 8 和图 10 可知，驱动器实际输出值与参考值的偏差逐渐趋于零，说明 PID 控制效果良好。

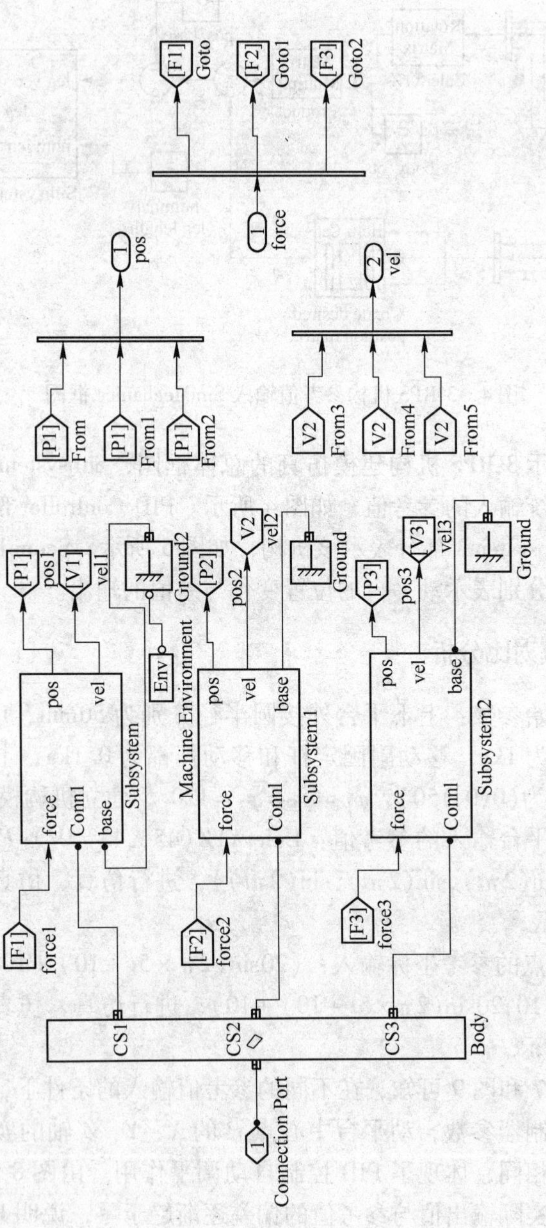

图 5 3-RPS 机构 SimMechanics 框图

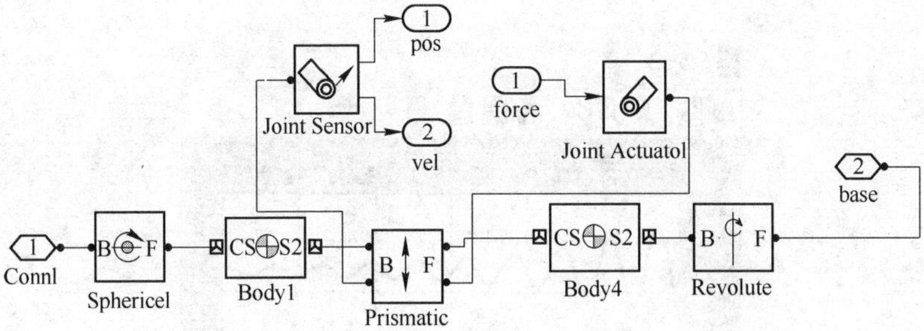

图 6　3-RPS 机构支链 SimMechanics 框图

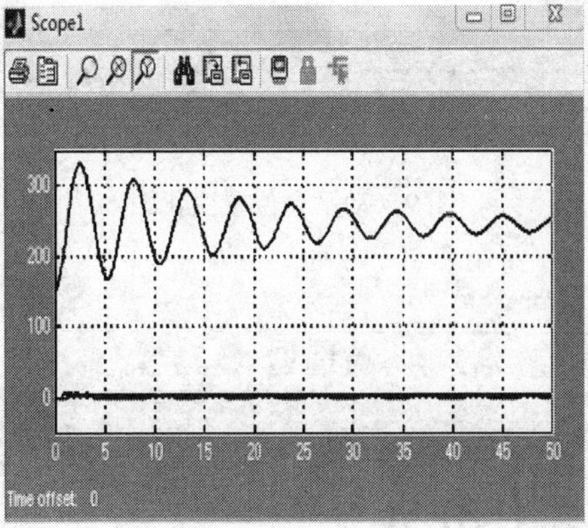

图 7　动平台位置变化图

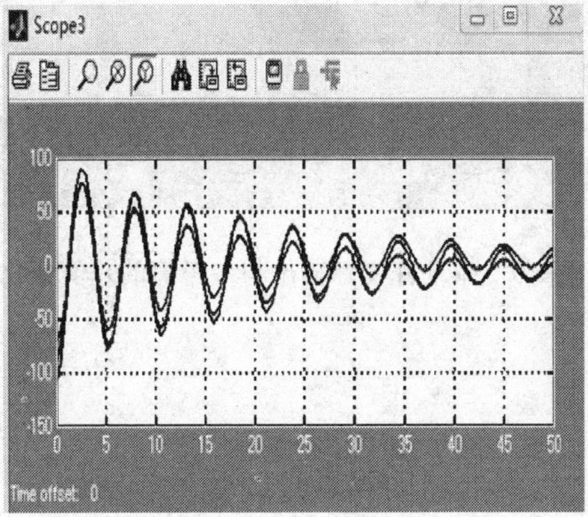

图 8　驱动杆实际输出与参考值的差值变化

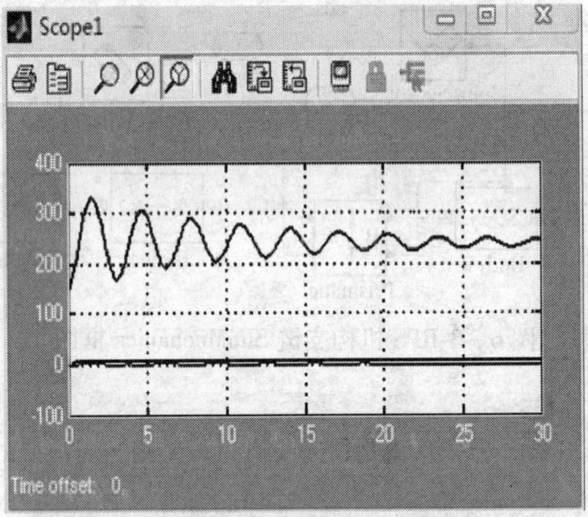

图 9 动平台位置变化图

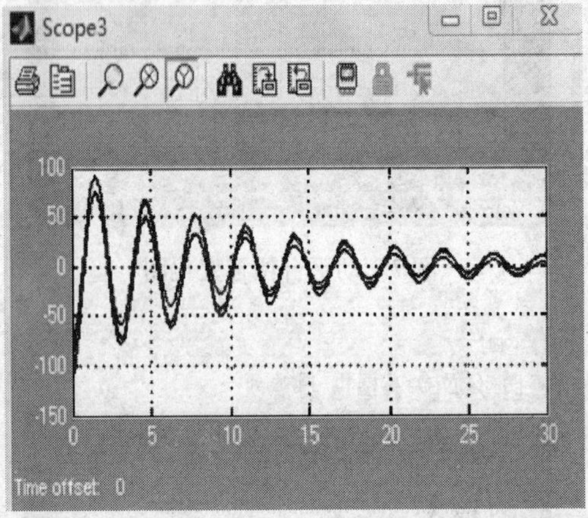

图 10 驱动杆实际输出与参考值的差值变化

2.2.6 结论

通过应用坐标变换法和位置封闭解法对 3-RPS 并联机器人的正反解及位置进行了分析，为进一步分析该机构运动学与控制仿真提供了有效的理论基础。根据控制原理，运用 Matlab 中 SimMechanics

模块，建立机构模型与控制仿真模型，在不同状态下进行仿真对比研究。仿真结果表明：基于PID控制可实现并联机器人动平台的自动调平功能，在实践中具有重要的应用价值。驱动器实际输出与参考值的误差值曲线在零点附近上下波动，逐渐趋于零，表明控制效果良好。

参 考 文 献

[1] 黄真，孔令富，方跃法. 并联机器人机构学理论及控制［M］. 北京：机械工业出版社，1997.

[2] 黄真，赵永生，赵铁石. 高等空间机构学［M］. 北京：高等教育出版社，2006.

[3] 朱大昌. 基于并联支撑机构的车载雷达天线自动调平系统研究［D］. 北京：北京交通大学，2008.

[4] 付廷贵，许瑛，杨光. 3自由度并联机构的位姿［J］. 南昌航空工业学院学报，2005，19（3）：25～31.

[5] 李树军，王阴，王晓光. 3-RPS并联机器人机构位置正解的杆长逼近法［J］. 东北大学学报，2001，22（3）：285～287.

[6] 吴巍. 一种两自由度并联机构位置分析与仿真［J］. 机械设计与制造，2011，5：113～115.

[7] 汪汇. 3－RRRT并联机器人运动仿真的SimMechanics实现［J］. 现代机械，2008，3：44～48.

[8] 刘胜，李晚龙，杜延春等. 并联机器人动力学与控制仿真研究［J］. 弹箭与制导学报，2005，25（7）：928～930.

[9] 王永超. 基于Matlab的机构运动仿真方法及其比较［J］. 计算机仿真，2004，21（8）：81～83.

[10] 梁毓明，陈德海. 轮式移动机器人调速系统的设计［J］. 江西理工大学学报，2008，29（4）：13～16.

2.3 Sliding Mode Synchronous Control for Fixture Clamps System Driven by Hydraulic Servo Systems

Although the two sides clips of a fixture clamps system have the same driving mechanisms, the synchronous error between the dual drivers of the two clips is generated by non-balanced forces. With position variation of clips and various uncertainty disturbances during the working process, the synchronous movement of the two clips is difficult. In the current paper, slidingmode synchronous controller is designed for fixture clamps system which is driven by hydraulic servo system. Setting the load dynamic error of one driver as external disturbance to the other driver, the departure from synchronization caused by the parameter variation and external forces between two clips is limited. The stability of the control system and the convergence of synchronous tracking errors are guaranteed by Lyapunov stability theory. Simulations illustrate the applicability of the proposed approach.

2. 3. 1 Introduction

Owing to rapid responses and high power-to-weight ratios, hydraulic servo system has been widely used in many industrial applications such as assembly tasks, mining, and material handling, etc. Recently, synchronization of hydraulic servo system and its application to fixture clamps have received much attention. Synchronous controller design for hydraulic manipulators, however, is challenging when compared with electrically actuated manipulators. In a hydraulic actuator, the control signal activates the spool valve which controls the flow of hydraulic fluid into and out of the actuator and fundamentally controls the derivative of the actuator force. Furthermore, hydraulic systems are highly non-linear and subject to parameter uncertainty, i. e. parameters change with time as a result of variations in operation condition and component degradation. In order to handle these situations, Schmidt et al. [1] proposed adaptive control of hydraulic systems with uncertainties. By using radial basis functions (RBF) networks to compensate for the effects of friction in the system, an adaptive controller is derived to control actuator force with unknown valve flow coefficients and fluid parameters. Zhu et al. [2] proposed an adaptive control scheme by using direct output force measurements through loadcells. In the proposed approach, the output force error resulting from direct measurement is used to update the parameters of a novel friction model that includes not only the coulomb-viscous friction force in sliding motion, but also the output force dependent friction force in presliding motion. Adaptive control schemes are widely used in hydraulic servo systems for uncertainties and highly non-linear characteristic[3~5].

Fuzzy and neural network control schemes have also been considered with the uncertainties of hydraulic servo system. Shao et al. [6] Considered the high nonlinearity and uncertain dynamics and presented a new non-linear hybrid controller composed of a classical proportional integral derivative (PID) controller and fuzzy controller based on self-adjusting modifying factor. Furthermore, a fuzzy switching mode was employed to avoid the undesirable disturbances caused by the switchover between the two control methods. Azimian et al. [7] Applied some neural networks based methods to solve a control benchmark problem by controlling an electro-hydraulic velocity servomotor system. As a popular non-linear control method, sliding mode control (SMC) has been considered to solve the non-linear friction existing in hydraulic systems. Wang et al. [8] applied SMC to a high-precision hydrostatic actuating system with nonlinear discontinuous friction. By defining a discrete time sliding surface for non-linear system, a linear quadratic approach was presented for the non-linear hydraulic servo system. SMC has been designed for the non-linear property and parameter variability of hydraulic servo system in references [9] and [10].

For the two clips moving in synchronization, the control problem combines with the classical SMC issue with the synchronous tracking control of the two hydraulic servo systems. In order to eliminate the synchronous error between the two hydraulic servo drivers, the control system established in this paper integrates the SMC and tracking control to perform the synchronous position of fixture clamps system. This paper is organized as follows. The modelling of the hydraulic servo system is described in section 1. The sliding mode synchronous control is presented to achieve the synchronization between two clips of the fixture clamps system. In addition, an important issue regarding the compensation of errors provided by two clips is discussed and several compensation schemes are reviewed in section 2. In section 3, the stability of the control system and the convergence of synchronous tracking errors are guaranteed by Lyapunov stability theory. Finally, the simulation results show the feasibility of the proposed control strategy.

2.3.2 Model of Hydraulic Servo Systems

The mathematical model of an asymmetric servovalve system for the hydraulic cylinder shown in Fig. 1 is as follows. Without any elastic load

$$Y(s)=\frac{(K_{x\alpha}/A_{me})x-\frac{K_{t\alpha}}{A_eA_{me}}\left(\frac{V_e}{4\beta_eK_{t\alpha}}s+1\right)f_e}{(s^2/\omega_{he}^2+2\xi_{he}/\omega_{he}+1)s} \tag{1}$$

where $K_{x\alpha}$ is the flow gain, m^2/s; A_{me} is the average area of piston, m^2; $K_{t\alpha}$ is the total flow-pressure coefficient, $m^2/(s\cdot N)$; A_e is the equivalent area of hydraulic cylinder, m^2; β_e is the modulus of elasticity, MPa; f_e is the equivalent outer interference force, N; x is the open size of valve, m; ω_{he} is the inherent frequency of hydraulic servo system; ξ_{he} is the damp ratio (Table 1).

$$\omega_{he}=2\sqrt{\frac{A_eA_{me}\beta_e}{V_em}}(\text{rad/s}),\ \xi_{he}=K_{t\alpha}\sqrt{\frac{m\beta_e}{V_eA_eA_{me}}}+\frac{B_p}{4}\sqrt{\frac{V_e}{m\beta_eA_eA_{me}}}$$

where m is the loading mass, kg; B_p is the viscidity damping coefficient of load, N·s.

It can be seen that with disturbing force the mathematical model of asymmetric hydraulic cylinder is very analogous to that of the symmetric one. When the valve spool is not in zero position and the piston moves in the direction of the chamber with rod, the movement of asymmetric cylinder can be regarded as a force exerted on symmetric cylinder load, which is equal to the pressure p_1 on the chamber without rod of asymmetric cylindermultiplied by rod area A_1. Likewise, when the piston moves to chamber without rod, the corresponding equivalent outer interference force can also be obtained. Therefore, system output is in fact decided by two inputs: displacement x of the valve spool and the equivalent outer interference force f_e.

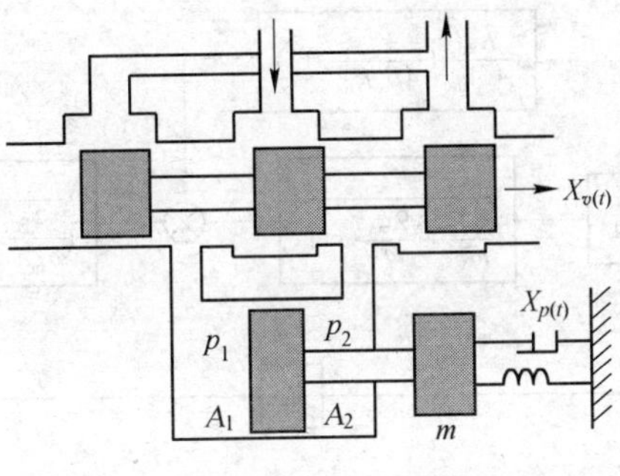

Fig. 1 Model of asymmetric servo valve

$$f_e = \begin{cases} \dfrac{\eta^2 A_1 p_1}{1+\eta^2} & x>0 \\ \dfrac{A_1 p_1}{1+\eta^2} & x<0 \end{cases}$$

where $\eta = A_2/A_1$.

Table 1 Actuator parameters

Parameter	Value
Pump pressure, p_s	8.259MPa
Exit pressure, p_e	0MPa
Valve constant, x	4.064×10^{-5}
Flow gain, $K_{x\alpha}$	$1.9568\times10^{-4}\,\mathrm{m/s^2}$
Bulk modulus of hydraulic fluid, β	568MPa
Position area, A_{me}	$3.167\times10^{-3}\,\mathrm{m^2}$
Equivalent area, A_e	$2.78\times10^{-3}\,\mathrm{m^2}$

The control system of the actuator is shown in Fig. 2. where K_q is the proportional coefficient of amplifier and motor loop and K_L is proportional coefficient of feedback of potentionmeter.

The uncertain load is symbolized by $\dot{T}_1(t,\theta_m,\dot{\theta}_m)$ which can be the external load. The state variable equation describing the dynamics of the hy-

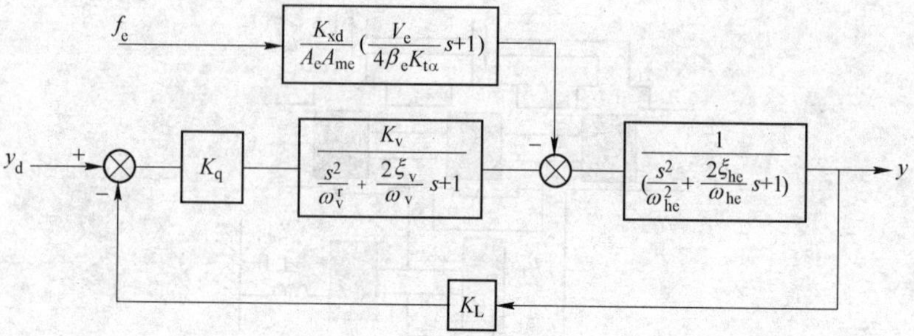

Fig. 2 Block diagram of control system of hydraulic actuator

draulic servo system is achieved as follows.

$$\dot{x}_1(t)=x_2(t),\ \dot{x}_2(t)=x_3(t),\ \dot{x}_3(t)=-a_1(t)x_1(t)-a_2(t)x_2(t)-a_3(t)x_3(t)+b(t)u(t)-d(t) \quad (2)$$

where

$$a_1(t)=\frac{4\beta_e K_{t\alpha} f_e}{V_e B_p},\ a_2(t)=\frac{f_e}{B_p}+\frac{4\beta_e K_{x\alpha}^2}{V_e B_p}+\frac{4\beta_e K_{t\alpha} B_m}{V_e B_p},\ a_3(t)=\frac{B_m}{B_p}+\frac{4\beta_e K_{t\alpha} f_e}{V_e}$$

$$b(t)=\frac{4\beta_e K_{x\alpha} B_m \xi_{ke}}{V_e B_p}>0,\ d(t)=\frac{4\beta_e K_{t\alpha} T_1(t,\theta_m,\dot{\theta}_m)}{V_e B_p}+\frac{\dot{T}_1(t,\theta_m,\dot{\theta}_m)}{B_p}$$

The linear parameterization of manipulator dynamics can be obtained in reference[11].

$$\boldsymbol{M}(\boldsymbol{q})\ddot{\boldsymbol{q}}+\boldsymbol{C}(\boldsymbol{q},\dot{\boldsymbol{q}})\dot{\boldsymbol{q}}+\boldsymbol{G}(\boldsymbol{q})+\boldsymbol{F}_{fr}+\boldsymbol{T}_d=\boldsymbol{JF} \quad (3)$$

where $\boldsymbol{q}$ is a vector of generalized joint position; $\boldsymbol{M}(\boldsymbol{q})$ is the clips inertia matrix; $\boldsymbol{C}(\boldsymbol{q},\dot{\boldsymbol{q}})$ is the Coriolis and centrifugal effects; $\boldsymbol{G}(\boldsymbol{q})$ is the gravitational term; F_{fr} is the friction force; $\boldsymbol{T}_d$ is the load disturbance; $\boldsymbol{F}$ is the vector of forces provided by the hydraulic actuators; $\boldsymbol{J}$ is the Jacobian from joint space to the linear actuator coordinate; $\boldsymbol{JF}$ is the torque originating from the actuator. Diagram of a hydraulic actuator with clips is shown in Fig. 3.

The matrices in equation (3) satisfy the following properties.

Property 1 $\boldsymbol{M}(\boldsymbol{q})$ is a symmetric positive definite matrix.

Property 2 $(\dot{\boldsymbol{M}}-2\boldsymbol{C})$ is a skew-symmetric matrix.

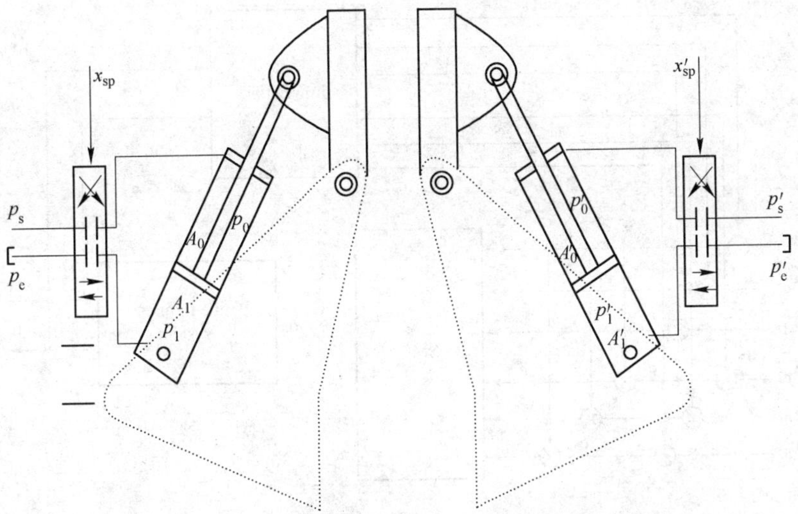

Fig. 3 Diagram of hydraulic actuators with clips

Property 3 $C(x,y)z = C(x,z)y \quad \forall x,y,z \in \mathbf{R}^{n\times 1}$.

Property 4 $\exists C_M$ s. t. $\|C(x,y)\| \leqslant C_M\|y\| \quad \forall x,y \in \mathbf{R}^{n\times 1}$.

In order to capture the mechanical uncertainties, those parameters appear in $M(q)$, $C(q,\dot{q})$, and $G(q)$, and the load disturbance T_d is assumed to be unknown positive constant.

The objective of this paper is to design a SMC for the hydraulic servo system of clamp subjecting to huge uncertainties and to keep the two hydraulic servo systems of clips synchronous.

2.3.3 The Sliding Mode Synchronous Control

The control system architecture is shown in Fig. 4.

In order to keep the hydraulic servo driver synchronous, the switching surface is considered to include the function $\Delta e_2 \to 0$.

Define the following switching surface

$$s_1 = c_0\ddot{e}_1(t) + c_1\dot{e}_1(t) + c_2 e_1(t) + c_3\int e(\tau)\mathrm{d}\tau \tag{4}$$

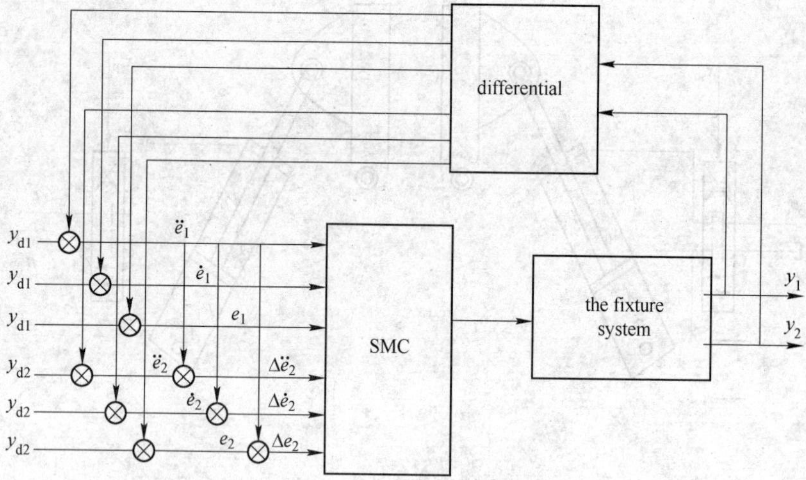

Fig. 4 Structure of sliding mode synchronous controller

$$s_2 = c_4\Delta\ddot{e}_2(t) + c_5\Delta\dot{e}_2(t) + c_6\Delta e_2(t) + c_7\int e(\tau)\mathrm{d}\tau \tag{5}$$

where $e_1(t) = y_{d1}(t) - y_1(t)$, $e_2(t) = y_{d2}(t) - y_2(t)$, $\Delta e_2(t) = e_2(t) - e_1(t)$.

c_i $(i=0\sim5)$ are chosen such that the dynamics of $\boldsymbol{s}=0$ is stable and has the desired eigenvalues. The assumptions of this paper are described as follows.

Assumption 1 The upper bound of parameter uncertainties is known.

Assumption 2 $|d(t)| < \varepsilon(t)$ indicates that an upper bound of uncertain load is known. However, if the system uncertainties are huge, the information of $a_i(t)$ $(i=1\sim3)$ is difficult to obtain because over-conservative design of the controller makes the system response oscillatory or even unstable[12~15]. If all the states are not available, and observer combined with a controller can be employed to deal with this kind of control problem[16].

Assumption 3 $\{\ddot{y}_{1(2)d}, \dot{y}_{1(2)d}, y_{1(2)d}\}$ are known, bounded and continuous.

Assumption 4 State vectors $x_i(t)$ are available.

Considering the hydraulic servo systems of two clips, control law can be composed of the equivalent control law u_{eq} and switching control law u_0, that is

$$u(t) = u_{eq}(t) + u_0(t) \tag{6}$$

with

$$\begin{aligned} u_{eq1}(t) &= \{c_0(\ddot{y}_{d1} + a_{11}(t)x_{11}(t) + a_{12}(t)x_{12}(t) + a_{13}(t)x_{13}(t) + \\ &\quad c_1(\ddot{y}_{d1} - x_{13}(t))) + c_2(\ddot{y}_{d1} - x_{12}(t)) + c_3(y_{d1} - x_{11}(t))\}/b_1(t) \\ u_{01} &= -K_1 \operatorname{sgn}(s_1) \end{aligned} \tag{7}$$

$$\begin{aligned} u_{eq2}(t) &= \{c_4[(\dddot{y}_{d1} + a_{11}(t)x_{11}(t) + a_{12}(t)x_{12}(t) + a_{13}(t)x_{13}(t)) - \\ &\quad (\dddot{y}_{d2} + a_{21}(t)x_{21}(t) + a_{22}(t)x_{22}(t) + a_{23}(t)x_{23}(t))] + \\ &\quad c_5[(\ddot{y}_{d1} - x_{13}(t)) - (\ddot{y}_{d2} - x_{23}(t))] + c_6[(\dot{y}_{d1} - x_{12}(t)) - \\ &\quad (\dot{y}_{d2} - x_{22}(t))] + c_7[(y_{d1} - x_{11}(t)) - (y_{d2} - x_{21}(t))]\}/b_2(t) \\ u_{02} &= -K_2 \operatorname{sgn}(s_2) \end{aligned} \tag{8}$$

2.3.4 Stability Analysis of Sliding Model Synchronous Controller

Suffering from disturbance and uncertainty, the system deviates from the manifold. The discontinous sliding model synchronous control must be designed to drive the system states back to the manifold and guarantee the stability. To design this control, a Lyapunov candidate is chosen as

$$\boldsymbol{V} = \frac{1}{2}\boldsymbol{s}_1^{\mathrm{T}}\boldsymbol{M}\boldsymbol{s}_1 + \frac{1}{2}\boldsymbol{s}_2^{\mathrm{T}}\boldsymbol{M}'\boldsymbol{s}_2 \tag{9}$$

Since $\boldsymbol{M}$ and $\boldsymbol{M}'$ are symmetrically and positively definite, it is obvious that $\boldsymbol{V}$ is a positive scalar function of vector $\boldsymbol{s}$ and time $\boldsymbol{t}$. The derivative of $\boldsymbol{V}$ is

$$\begin{aligned} \dot{V} &= \frac{1}{2}(\dot{\boldsymbol{s}}_1^{\mathrm{T}}\boldsymbol{M}\boldsymbol{s}_1 + \boldsymbol{s}_1^{\mathrm{T}}\dot{\boldsymbol{M}}\boldsymbol{s}_1 + \boldsymbol{s}_1^{\mathrm{T}}\boldsymbol{M}\dot{\boldsymbol{s}}_1) + \\ &\quad \frac{1}{2}(\dot{\boldsymbol{s}}_2^{\mathrm{T}}\boldsymbol{M}\boldsymbol{s}_2 + \boldsymbol{s}_2^{\mathrm{T}}\dot{\boldsymbol{M}}\boldsymbol{s}_2 + \boldsymbol{s}_2^{\mathrm{T}}\boldsymbol{M}\dot{\boldsymbol{s}}_2) \end{aligned} \tag{10}$$

Since Property 1 of the dynamics gives

$$\dot{\boldsymbol{s}}_1^{\mathrm{T}}\boldsymbol{M}\boldsymbol{s}_1 = \boldsymbol{s}_1^{\mathrm{T}}\boldsymbol{M}\dot{\boldsymbol{s}}_1, \quad \dot{\boldsymbol{s}}_2^{\mathrm{T}}\boldsymbol{M}\boldsymbol{s}_2 = \boldsymbol{s}_2^{\mathrm{T}}\boldsymbol{M}\dot{\boldsymbol{s}}_2$$

The following can be obtained

$$\dot{\boldsymbol{V}} = \frac{1}{2}(\boldsymbol{s}_1^{\mathrm{T}}\dot{\boldsymbol{M}}\boldsymbol{s}_1 + \boldsymbol{s}_2^{\mathrm{T}}\dot{\boldsymbol{M}}'\boldsymbol{s}_2) + \boldsymbol{s}_1^{\mathrm{T}}\boldsymbol{M}\dot{\boldsymbol{s}}_1 + \boldsymbol{s}_2^{\mathrm{T}}\boldsymbol{M}'\dot{\boldsymbol{s}}_2 \tag{11}$$

Using equations (4) to (8), $\boldsymbol{M}\dot{\boldsymbol{s}}_1$ and $\boldsymbol{M}'\dot{\boldsymbol{s}}_2$ can be denoted as

$$\boldsymbol{M}\dot{\boldsymbol{s}}_1 = u_1 - u_{eq1},\ \boldsymbol{M}'\dot{\boldsymbol{s}}_2 = u_2 - u_{eq2}$$

Since Property 2 of the dynamics gives

$$\boldsymbol{s}_1^{\mathrm{T}}(\dot{\boldsymbol{M}} - 2\boldsymbol{C})\boldsymbol{s}_1,\ \boldsymbol{s}_2^{\mathrm{T}}(\dot{\boldsymbol{M}}' - 2\boldsymbol{C})\boldsymbol{s}_2$$

Then equation (9) can be expressed as

$$\begin{aligned}\dot{V} &= \boldsymbol{s}_1^{\mathrm{T}}(-K_1\operatorname{sgn}(\boldsymbol{s}_1) + \boldsymbol{C}\boldsymbol{s}_1 - u_{eq1}) + \boldsymbol{s}_2^{\mathrm{T}}(-K_2\operatorname{sgn}(\boldsymbol{s}_2) + \boldsymbol{C}\boldsymbol{s}_2 - u_{eq2}) \\ &\leqslant \boldsymbol{s}_1^{\mathrm{T}}(-K_1\operatorname{sgn}(\boldsymbol{s}_1) + |\boldsymbol{C}\boldsymbol{s}_1 - \boldsymbol{u}_{eq1}|) + \boldsymbol{s}_2^{\mathrm{T}}(-K_2\operatorname{sgn}(\boldsymbol{s}_2) + \\ &\quad |\boldsymbol{C}\boldsymbol{s}_2 - \boldsymbol{u}_{eq2}|)\end{aligned} \tag{12}$$

Design

$$K_1 > |\boldsymbol{C}\boldsymbol{s}_1 - u_{eq1}|,\ K_2 > |\boldsymbol{C}\boldsymbol{s}_2 - u_{eq2}|$$

Thus, if $\boldsymbol{s}_1 > 0$, $\boldsymbol{s}_2 > 0$, then $\operatorname{sgn}(\boldsymbol{s}_1) = \operatorname{sgn}(\boldsymbol{s}_2) = 1$, from equations (10) and (11)

$$-K_1\operatorname{sgn}(\boldsymbol{s}_1) + |\boldsymbol{C}\boldsymbol{s}_1 - u_{eq1}| < 0,\ -K_2\operatorname{sgn}(\boldsymbol{s}_2) + |\boldsymbol{C}\boldsymbol{s}_2 - u_{eq2}| < 0$$

Therefore

$$\boldsymbol{s}_1^{\mathrm{T}}(-K_1\operatorname{sgn}(\boldsymbol{s}_1) + |\boldsymbol{C}\boldsymbol{s}_1 - u_{eq1}|) + \boldsymbol{s}_2^{\mathrm{T}}(-K_2\operatorname{sgn}(\boldsymbol{s}_2) + |\boldsymbol{C}\boldsymbol{s}_2 - u_{eq2}|) < 0 \tag{13}$$

On the other hand, if $\boldsymbol{s}_1 < 0$, $\boldsymbol{s}_2 < 0$, then $\operatorname{sgn}(\boldsymbol{s}_1) = \operatorname{sgn}(\boldsymbol{s}_2) = -1$, from equations (10) and (11)

$$-K_1\operatorname{sgn}(\boldsymbol{s}_1) + |\boldsymbol{C}\boldsymbol{s}_1 - u_{eq1}| > 0,\ -K_2\operatorname{sgn}(\boldsymbol{s}_2) + |\boldsymbol{C}\boldsymbol{s}_2 - u_{eq2}| > 0$$

Therefore

$$\boldsymbol{s}_1^{\mathrm{T}}(-K_1\operatorname{sgn}(\boldsymbol{s}_1) + |\boldsymbol{C}\boldsymbol{s}_1 - u_{eq1}|) + \boldsymbol{s}_2^{\mathrm{T}}(-K_2\operatorname{sgn}(\boldsymbol{s}_2) + |\boldsymbol{C}\boldsymbol{s}_2 - u_{eq2}|) < 0 \tag{14}$$

Substituting equations (13) and (14) into equation (13) the following can be obtained

$$\dot{V} < 0 \tag{15}$$

According to Lyapunov's direct method, the provided system is stable in the sense of Lyapunov. The sliding model synchronous control is applied by using a large enough switch gain K_1 and K_2, respectively, and by shifting the sign according to different states in motion to force the system to be passive (Table 2). The stability is guaranteed under the assumption that, K_1

and K_2 can be determined conservatively.

Table 2 Condition of reference trace with square wave and sine wave, respectively

Reference type	Amplitude	Frequency	Sample time
Square wave	0.5mm	0.34s	0.1s
Sine Wave	1mm	1s	0.1s

Trace tracking of the selected standard driver and the other driver is shown in Fig. 7 and Fig. 8 respectively.

2.3.5 Simulations

The non-linear sliding model synchronous controller presented previously is tested by numerical simulations to verify the stability and synchronously tracking performance. The parameters of actuators are given as follows.

Although the load disturbances were assumed unknown constants, in order to investigate the control performance, they are changed as follows: load disturbances of $T_d = 400\text{N/m}$ are injected into the clamp system at $t = 4\text{s}$ and removed out at $t = 8\text{s}$. The diagonal values of those diagonal control gains are chosen as follows: $K_1 = 6000 \sim 10000$, $K_2 = 700 \sim 1000$.

The error tracking of the clip selected as a standard is shown in Fig. 5 and Fig. 6.

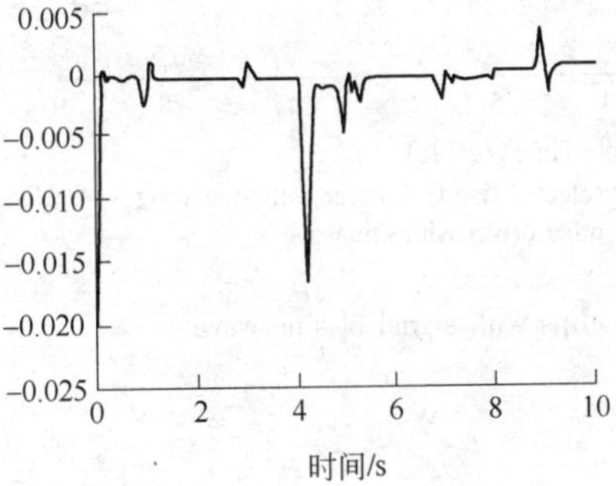

Fig. 5 Error tracking of the clip selected as a standard

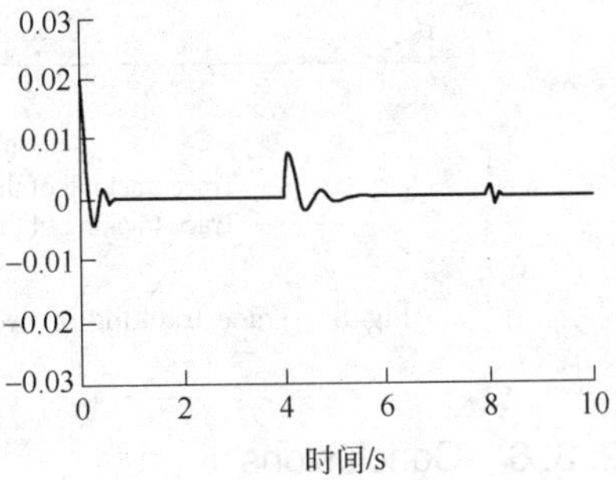

Fig. 6 Synchronous error between two clips

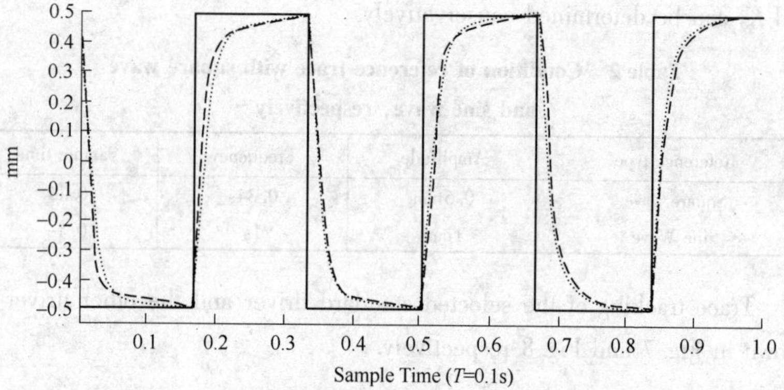

Fig. 7 Trace tracking of two clips with signal of square wave

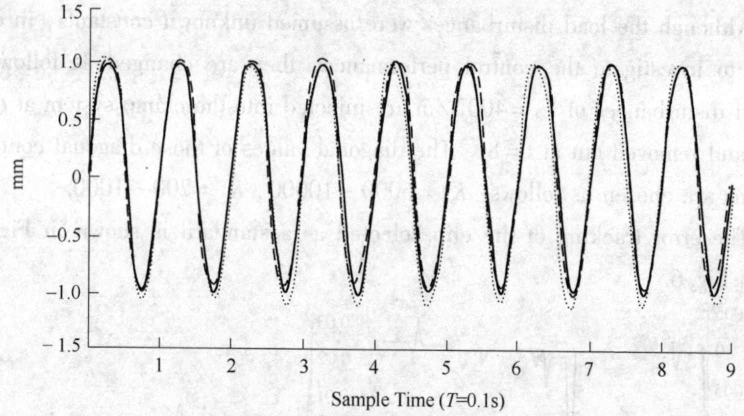

Fig. 8 Trace tracking of two clips with signal of sine wave

2.3.6 Conclusions

This paper proposes a novel method for the fixture clamps system with sliding model synchronous control strategy. As for the multi-input (more than two inputs) system, in order to keep two or more actuators synchronous, one actuator is selected as a standard. A sliding model synchronous controller can be designed based on the synchronous errors which are obtained by comparing the output signals of other actuators with that of the selected standardized actuator. Under some conditions, the proposed controller is accomplished by Lyapunov stability theory and the proposed scheme indeed improves the synchronous of the clamp system. The simulation results confirm that the developed method is useful for controlling a class of multi-variable systems with the demand of synchronous control.

References

[1] Schmidt D F, Paplinski A B, Lowe G S. Adaptive control of hydraulic systems with MML inferrec RBF networks [C] //In Proceedings of the 2005 IEEE International Conference on Robotics and Automation, Barcelona, Spain, 2005: 2362 ~ 2375.

[2] Zhu W H, Dupuis E, Piedboeuf J C. Adaptive output force tracking control of hydraulic cylinders [C] //In Proceedings of the 2004 American Control Conference, Boston, MA, 2004: 970 ~ 976.

[3] Guan C and Zhu S. Adaptive time-varying sliding mode control for hydraulic servo system [C] //2004 8th International Conference on Control, Automation, Robotics and Vision, Kunming, China, 2004: 1774 ~ 1780.

[4] Zhu X, Wang H, Zhao M, Zhou J. An alterable gain adaptive control compensation methods for hydraulic servo joint based on ANFIS [C] //In Proceedings of the IEEE International Conference on Mechatronics and Automation, Niagara Falls, Canada, 2005: 148 ~ 155.

[5] Garagic D and Srinivasan K. Application of non-linear adaptive control techniques to an electrohydraulic velocity servomechanism [J]. IEEE Trans Control Syst. Technol., 2004, 12 (2): 303 ~ 315.

[6] Shao J, Chen L, Ji Y and Sun Z. The application of fuzzy control strategy in electro-hydraulic servo sytem [C] //In Proceedings of ISCIT 2005, 2005: 159 ~ 165.

[7] Azimian H, Adlgostar R, Teshnehlab M. Velocity control of an electro hydraulic servomotor by neural networks [C] //In Proceedings of International Conference on Physics and Control, St Petersburg, Russia, 2005: 677 ~ 682.

[8] Wang S, Habibi S, Burton R, Sampson E. Sliding mode control for a model of an electrohydraulic actuator system with discontinuous non-linear friction [C] //In Proceedings of the 2006 American Control Conference, Minneapolis, Minnesota, USA, 2006: 5897 ~ 5905.

[9] Wang H, Zhao K, Cui X. Sliding control approach to pneumatic hydraulic combination control

(PHCC) servo system [C] //In Proceedings of the 6th World Congress on Intelligent Control and Automation, Dalian, China, 21 ~ 23 June, 2006: 2156 ~ 2160.

[10] Perron M, de Lafontaine J, Desjardins Y. Sliding mode control of a servomotor-pump in a position control application. In Proceedings of IEEE CCECE/CCGEI, Saskatoon, 2005: 1287 ~ 1292.

[11] Zheng H, Sepehri N. Adaptive backstepping control of hydraulic manipulators with friction compensation using LuGre model [C] //In Proceedings of the 2006 American Control Conference, Minneapolis, Minnesota, USA, 2006: 3164 ~ 3170.

[12] Harashima F, Hashimoto H, Kondo S. MOSFET converter-fed position servo system with sliding mode control [J]. IEEE Trans. Ind. Electron., 1985, 32 (3): 238 ~ 244.

[13] Slotin J E, Coetsee J A. Adaptive sliding controller synthesis for non-linear systems [J]. Int. J. Control, 1986, 43 (6): 1631 ~ 1651.

[14] Kachroo P, Tomizuka M. Chattering reduction and error convergence in the sliding mode control of a class of non-linear systems [J]. IEEE Trans. Autom. Control, 1996, 32 (7): 1063 ~ 1068.

[15] Oucheriah S. Robust sliding mode control of uncertain dynamic delay systems in the presence of matched and unmatched uncertainties [J]. ASME J. Dyn. Syst. Meas, Control, 1997: 119, 69 ~ 72.

[16] Hwang C L, Sung F Y. Neuro-observer controller design for non-linear dynamical systems [C] //In Proceedings of 35th IEEE Conference on Decision and Control, Kobe, Japan, 1996: 3310 ~ 3316.

2.4 Neural-adaptive Sliding Mode Control of 4-SPS (PS) Type Parallel Manipulator

This paper presents a neural-adaptive sliding mode control for the tracking control of 4-SPS (PS) type parallel manipulator. The neural-adaptive controller is introduced to modify the coefficients of sliding manifold in sliding control strategy, which solve the problem that the equivalent control can not be obtained accurately because of the uncertain and fixed coefficients of sliding manifold and external disturbances of the system. So the controller can be designed without depend on fixed sliding manifold as general design method. Accordingly, it can be more effectively to solve the chattering in sliding mode control. The nonlinear controller, which guarantees the stability of the proposed control system based on Lyapunov stability theory, is also developed. Simulation results show that the control approach can decrease the tracking error, enhance the system's robustness and restrain the chattering effectively in the sliding mode control.

2.4.1 Introduction

Early research on parallel manipulators concentrated mainly on six-degrees-of-freedom (6-DoF) Steward-Gough type manipulators[1]. However, a 6 DoF fully parallel manipulator has the limitations of small workspace, difficult direct kinematics, and complex mechanical design. For some industrial applications, a parallel manipulator with fewer than 6 DoF, called a low-DoF parallel manipulator, is sufficient. However, the Jocabian matrix of low DoF parallel manipulators is not phalanx which is differcult to get the kinetic equation for control design. On the other hand, parallel manipulators are typical nonlinear system with uncertain structure and variable parameters. So it is unsuitable design the control system for parallel manipulator with linear control theory. Much effort has been focussed on the trajectory tracking problem applied to parallel manipulators. Park et al. [2] proposed the sliding mode controller with perturbation observe for 6-DoF parallel manipulator in the presence of nonlinear and uncertainty terms. Li et al. [3] also focused on the uncertainty of the complex robot sysem and proposted a novel control schem for 6 DoF parallel manipulators. This control strategy combined cascaded Cerebellar Model Articulation Controller (CMAC) with variable structure control (VSC). Where the VSC is used to reduce the effect of CMAC estimate error and unrepeatable disturbances. George and Brian[4] designed a three dimensional fuzzy PID controller to a 6-DoF Steward-Gough based parallel manipulator. The proposed controller is performed without using the forward kinematics of the manipulator. Wu and Handroos[5] combined the fuzzy self-tuning PD and the fuzzy self-tuning PI by using a simple design scheme for a 6-DoF parallel manipulator. The essential idea of the control scheme is to take the advantage of the fuzzy self-tuning PD and PI controller. Ouyang et al. [6] presented a study of examing nonlinear PD (NPD) control of muti-DoF parallel manipulator systems for a generic task. They were also examine how the mechanical structure of the manipulator affects dynamic performance. Yiu and Li[7] presented a PID and a model-based adaptive robust controllers for a 2-DoF parallel manipulator with actuator redundancy and implement simple friction models to compensate the frictions in the active joints, passive joints, as well as the multiple joint. Su et al. [8] proposed a simple synchronized control algorithm by incorporating cross-coupling technolgy into a common PD control architecture for the control of parallel manipulators. By synchronizing all actuators' motions, the differential position errors amongst actuators converge to zero.

In this paper, sliding model control strategy based on neuro-adaptive is presented for 4-SPS (PS) type parallel manipulator. The new control synthesis strategy has two components. First, a sliding model controller is designed based on a referenced model which forced the tracking errors to the sliding face where the errors estimate to zero. Next, a neuro-adaptive controller is designed, which trains online to model the uncertainties caused by parameters variations, friction, external load. Note that the neural network training algorithm is derived based on Lyapunov stability theory, which guarantees both stability of the error dynamics as well as boundedness of the weights. This paper is organized as follows. In section 2, motion characteristic of 4-SPS (PS) type parallel manipulator is analyzed in detail by screw theory. In section 3, the design and analysis of the controller is presented. The proposed controller and its stability analysis are presented in section 4. Simulation results of 4-SPS (PS) type parallel manipulator are given in section 5 and followed by conclusion in section 6.

2.4.2 Motion Characteristic and Dynamic of 4-SPS (PS) Type Parallel Manipulator

2.4.2.1 Motion characteristic of 4-SPS (PS) type parallel manipulator

A schematic representation of the 4-SPS (PS) type parallel manipulator is shown in Fig. 1, where the fixed plate is labeled as FP and the moving platform is MP. Indeed MP is connected to FP through four identical leg mechanisms, and is driven by the corresponding hydraulic servo actuators. An especial leg which only provides restrict to MP connected MP to FP in the middle of each plate. A prismatic joint connected MP by a spherical joint and another side is vertical fixed on FP.

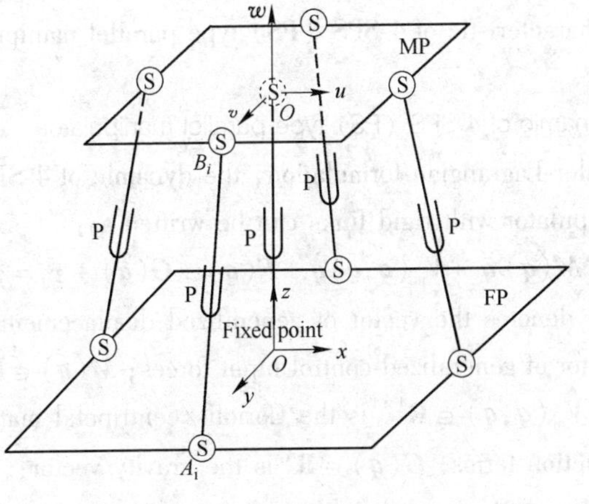

Fig. 1 Kinematics architecture and design parameters of 4-SPS (PS) type parallel manipulator

Particularly, links of the ith leg of the mechanism, are identified through d_i, which is the length of the frame link. The kinematics variables are Δd_i, which is the input distant. Finally, the size of MP and FP are given by a and b respectively. $o'$is the center point of MP, o is the center point of FP. Based on screw theory[9,10], the restrict provided by center limb can be denoted as follows.

Assume that one direction of the spherical joint is $\boldsymbol{s}_1 = (n, m, o)$, the two other directions of the spherical are $\boldsymbol{s}_2 = \left(-\frac{o}{n}, 0, 1\right)$ and $\boldsymbol{s}_3 = \left(-\frac{m}{n}, 1, 0\right)$.

Screw system of the center limb is

$$\$_1=(0,0,0,0,0,1),\ \$_2=(n,m,o,-mz,nz,1),$$

$$\$_3=\left(-\frac{o}{n},0,1,0,-\frac{zo}{n},1\right),\ \$_4=\left(-\frac{m}{n},1,0,-z,-\frac{zm}{n},0\right)$$

The reciprocal screw can be denoted as following.

$$\hat{\$}_1=(0,1,0,-z,0,0),\ \hat{\$}_2=(1,0,0,0,z,0) \tag{1}$$

It denotes that two pure force acted on the MP, the acting point at o′, so the motion characteristic of 4-SPS (PS) type parallel manipulator is 3R1T kinetic form.

2.4.2.2 Dynamic of 4-SPS (PS) type parallel manipulator

Using the Euler-Lagrangian formulation, the dynamic of 4-SPS (PS) type parallel manipulator with rigid links can be written as:

$$\boldsymbol{M}(\boldsymbol{q})\ddot{\boldsymbol{q}}+V_{\mathrm{m}}(\boldsymbol{q},\dot{\boldsymbol{q}})\dot{\boldsymbol{q}}+\boldsymbol{F}(\dot{\boldsymbol{q}})+\boldsymbol{G}(\boldsymbol{q})+\tau_{\mathrm{d}}=\tau \tag{2}$$

where $\boldsymbol{q}\in\mathbf{R}^4$ denotes the vector of generalized displacements; $\tau\in\mathbf{R}^4$ denotes the vector of generalized control input forces; $\boldsymbol{M}(\boldsymbol{q})\in\mathbf{R}^{4\times4}$ is the inertia matrix; $V_{\mathrm{m}}(\boldsymbol{q},\dot{\boldsymbol{q}})\in\mathbf{R}^{4\times4}$ is the Coriolis/centripetal matrix; $\boldsymbol{F}(\dot{\boldsymbol{q}})\in\mathbf{R}^4$ are the friction terms; $\boldsymbol{G}(\boldsymbol{q})\in\mathbf{R}^4$ is the gravity vector; $\tau_{\mathrm{d}}\in\mathbf{R}^4$ represents disturbances which are bounded. Some fundamental properties of robot dynamics are in reference[11].

Property 1 The inertia matrix $\boldsymbol{M}(\boldsymbol{q})$ is symmetric, uniformly positive definite, and bounded above and below so that

$$0<\lambda_1\boldsymbol{I}\leqslant\boldsymbol{M}(\boldsymbol{q})\leqslant\lambda_2\boldsymbol{I} \tag{3}$$

where $\boldsymbol{I}$ is the 4×4 identity matrix, λ_1 and λ_2 are positivescalar constants.

Property 2 The Coriolis/centripetal matrix can always be selected so that the matrix $[\boldsymbol{M}(\boldsymbol{q})-2V_{\mathrm{m}}(\boldsymbol{q},\dot{\boldsymbol{q}})]$ is skew symmetric.

The control objective can be stated as follows: given desired trajectories $\boldsymbol{q}_{\mathrm{d}}$、$\dot{\boldsymbol{q}}_{\mathrm{d}}$、$\ddot{\boldsymbol{q}}_{\mathrm{d}}\in\mathbf{R}^4$ which are bounded function of time, determine a equivalent control law $\tau_{\mathrm{eq}}(t)$, in the presence of parameter and other uncertainties such that the tracking error $e(t)\in\mathbf{R}^4$ and its derivatives tend to zero asymptotically, where

$$e(t)=\boldsymbol{q}_{\mathrm{d}}(t)-\boldsymbol{q}(t) \tag{4}$$

$$\dot{e}(t) = \dot{\boldsymbol{q}}_{\mathrm{d}}(t) - \dot{\boldsymbol{q}}(t) + \Delta \tag{5}$$

$$\ddot{e}(t) = \ddot{\boldsymbol{q}}_{d}(t) - \ddot{\boldsymbol{q}}(t) + \Delta' \tag{6}$$

where Δ and Δ' denote the nonlinear higher order of $\dot{\boldsymbol{q}}_{\mathrm{d}}(t) - \dot{\boldsymbol{q}}(t)$ and $\ddot{\boldsymbol{q}}_{\mathrm{d}}(t) - \ddot{\boldsymbol{q}}(t)$ by Fourier expanded equation respectively.

The tracking control problem in this paper is studied under the following assumptions:

Assumption 1 The external disturbance is limited by the following inequality:

$$\tau_{\mathrm{di}}(\boldsymbol{q}_i, \dot{\boldsymbol{q}}_i, t) \leqslant \max(c_0 + c_1 \|\boldsymbol{q}_i\| + c_2 \|\dot{\boldsymbol{q}}_i\| + c_3 \|\boldsymbol{q}_i\|^2 + c_4 \|\dot{\boldsymbol{q}}_i\|^2) \tag{7}$$

where $i = 1 \sim 4$, c_j $(j = 1, 2, \cdots, 4)$ is unknown constant matrix.

Assumption 2 The desired trajectories $\boldsymbol{q}_{\mathrm{di}}$、$\dot{\boldsymbol{q}}_{di}$ and $\ddot{\boldsymbol{q}}_{di}$ are bounded and given by the perfect model of 4-SPS (PS) type parallel manipulator.

2.4.3 Neural-adaptive Sliding Mode Controller

2.4.3.1 Traditional sliding mode controller

The design procedure of the SMC is a two-stage process[12~14]. The first phase is to choose a switching surface which is stable and has a desired behavior. The second phase is to determine a control law that forces the system's trajectory into the neighborhood of switching surface satisfying some conditions such that an asymptotical tracking can be guaranteed.

The switching function is defined as follows.

$$\boldsymbol{s} = \boldsymbol{w}\boldsymbol{e} + \dot{\boldsymbol{e}} \tag{8}$$

Based on proportional switching algorithm, the control law is

$$u_{\mathrm{eq}} = (\alpha |e| + \beta \dot{e}) \operatorname{sgn}(s) \tag{9}$$

where α and β are non-negative constants.

Set the state variable denoted as the errors of displacement and velocity of actuator joints, the nonlinear state equation of (3) is

$$\dot{x} = f(e_i, \dot{e}_i, \ddot{e}_i, t) + g(e_i, \dot{e}_i, t)u + \varphi(e_i, \dot{e}_i, t) \tag{10}$$

$$y = h(e_i, \dot{e}_i, t)x + \varphi(e_i, \dot{e}_i, t)u \tag{11}$$

The sliding surface is

$$s(x)=0 \tag{12}$$

The equivalent sliding model control law can be given by Eq.(8)~Eq.(12).

2.4.3.2 Neural-adaptive algorithm applied to sliding mode controller

In conventional SMC, the coefficient w in (9) is designed to constant. Actually, how to select these coefficients is more important to a nonlinear uncertainties system. The structure of SMC with neural-adaptive algorithm is shown in Fig. 2.

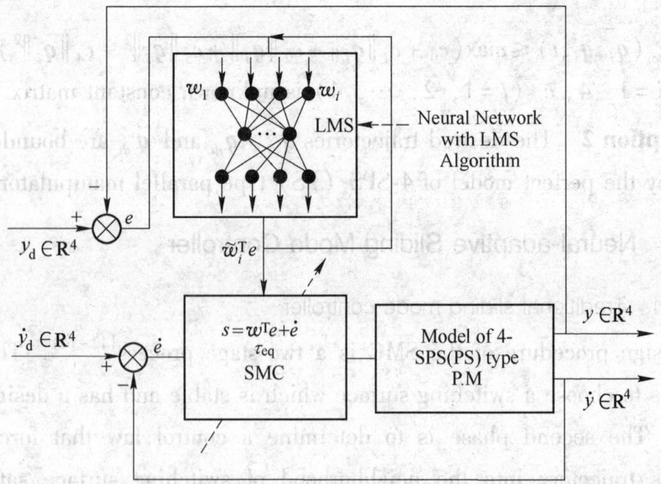

Fig. 2 The structure of SMC with neural-adaptive algorithm

To find the weight updates the error for weights should be found. After finding the error weights are updated by LMS algorithm as

$$\tilde{\boldsymbol{w}}_{ij}=\boldsymbol{w}_{ij}-\eta\frac{\mathrm{d}E}{\mathrm{d}w_{ij}} \tag{13}$$

where $\boldsymbol{w}_{ij}$ connects the jth node to i^{th} node.

The main working principle of the NN-LMS controller is to minimize a given error function by weights switching of coefficient $\boldsymbol{w}_{ij}$, which in turn pushes the error of the system to go to zero.

A Lyapunov function is defined in the following.

$$V = \frac{1}{2} s^{\mathrm{T}} s \tag{14}$$

$$\frac{\mathrm{d}V}{\mathrm{d}t} = -s^{\mathrm{T}} G s \tag{15}$$

where G is a positive definite matrix.

In order to achieve

$$\dot{s} + Gs = 0 \tag{16}$$

LMS algorithm is done by introduction the following error function to the NN controller.

$$E = \frac{1}{2}(\dot{s} + Gs)^{\mathrm{T}}(\dot{s} + Gs) \tag{17}$$

The errors can be calculated as shown in the following equations

$$\frac{\partial E}{\partial w_{ij}} = (\dot{s} + Gs)^{\mathrm{T}} \frac{\partial(\dot{s} + Gs)}{\partial w_{ij}} \tag{18}$$

$$\frac{\partial E}{\partial w_{ij}} = (\dot{s} + Gs)^{\mathrm{T}} D \frac{\partial(\dot{x})}{\partial u_{\mathrm{eq}}} \frac{\partial u_{\mathrm{eq}}}{\partial w_{ij}} \tag{19}$$

Following the definition, $\delta_i = (s + Gs)^{\mathrm{T}}$ DB, weight updates for the hidden layer can be found as

$$\frac{\partial E}{\partial w_{ij}} = \sum_k (\delta_k w_{ij}) f'(net_i)_B \tag{20}$$

2.4.4 Stability Analysis of Controller

System stability is analyzed by use of the Lyapunov function.

$$V = \frac{1}{2} s^{\mathrm{T}} s + \frac{1}{2\eta} \tilde{w}_{ij}^{\mathrm{T}} \tilde{w}_{ij} \tag{21}$$

Case one For $|s| < \varepsilon$, (21) implies

$$V = \frac{1}{2\eta} \tilde{w}_{ij}^{\mathrm{T}} \tilde{w}_{ij} > 0 \tag{22}$$

Hence, V is a suitable, positive-definite Lyapunov function candidate for the given range of e_i.

Differentiation with respect to time gives

$$\dot{V} = \frac{1}{\eta} \tilde{w}_{ij}^{\mathrm{T}} G' \frac{\mathrm{d}\tilde{w}_{ij}}{\mathrm{d}t} \tag{23}$$

where G' is a positive definite matrix. Being a dynamic structure network, $\frac{\mathrm{d}\,\tilde{\boldsymbol{w}}_{ij}}{\mathrm{d}t} \cong 0$, hence

$$\dot{V} = 0 \tag{24}$$

Case two For $|\boldsymbol{s}| \geqslant \varepsilon$, (21) implies

$$V = \frac{1}{2}\boldsymbol{s}^{\mathrm{T}}\boldsymbol{s} + \frac{1}{2\eta}\tilde{\boldsymbol{w}}_{ij}^{\mathrm{T}}\tilde{\boldsymbol{w}}_{ij} > 0 \tag{25}$$

Hence, $\boldsymbol{V}$ is a suitable, positive-definite Lyapunov function candidate. Differentiation with respect to time gives

$$\dot{V} = -\boldsymbol{s}^{\mathrm{T}}\boldsymbol{G}\boldsymbol{s} + \frac{1}{\eta}\tilde{\boldsymbol{w}}_{ij}^{\mathrm{T}}\boldsymbol{G}'\frac{\mathrm{d}\,\tilde{\boldsymbol{w}}_{ij}}{\mathrm{d}t} = -(\tilde{\boldsymbol{w}}_{ij}e + \dot{e})^{\mathrm{T}}\boldsymbol{G}(\tilde{\boldsymbol{w}}_{ij}e + \dot{e}) + \tilde{\boldsymbol{w}}_{ij}\left(\frac{1}{\eta}\frac{\mathrm{d}\tilde{\boldsymbol{w}}_{ij}}{\mathrm{d}t} - sf(net_i)\right) \tag{26}$$

Choosing

$$(\tilde{\boldsymbol{w}}_{ij}e + \dot{e})f(net_i) \geqslant \frac{1}{\eta}\frac{\mathrm{d}\tilde{\boldsymbol{w}}_{ij}}{\mathrm{d}t} \tag{27}$$

Implies $\dot{V} \leqslant 0$.

2.4.5 Simulations

For the Neural-adaptive SMC controller derived in the previous section, we investigate the performance of the system under practical environmental conditions in order to show the effectiveness of the proposed controller. 4-SPS (PS) type parallel manipulator is considered. The initial size of the 4-SPS (PS) type parallel manipulator is selected as follows.

$a = 100$mm; $b = 80$mm; $h = 40$mm; Force of 1# and 2# Limb are 20N

Force of 3# is 10N, Force of 4# is 15N

Trajectory of MP is shown in Fig. 3.

Trajectory of limb (1 ~ 4) are shown in Fig. 4 ~ Fig. 7 respectively.

2.4.6 Conclusions

In this paper, we have combined the concepts of neural-adaptive algo-

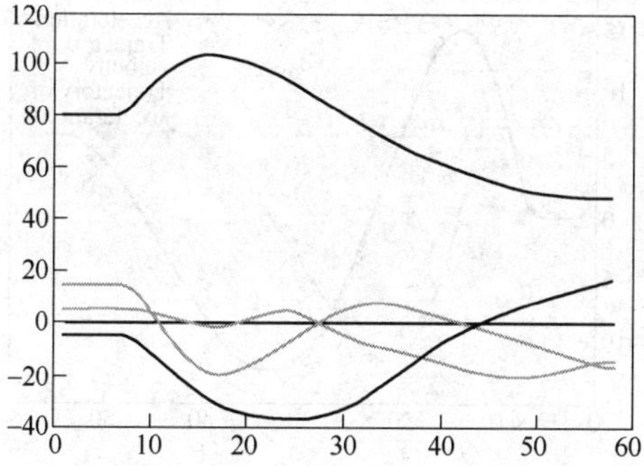

Fig. 3 Trajectory of motion platform

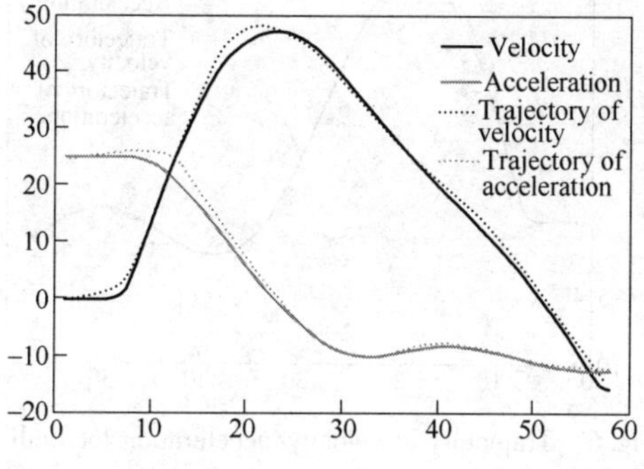

Fig. 4 Trajectory of velocity/acceleration for limb1#

rithm, SMC control scheme for motion control of the 4-SPS (PS) type parallel manipulator. In order to overcome the uncertainties and external disturbances which bring chattering to SMC controller, Neural-adaptive algorithm is used to modify the coefficient of sliding switch function. The proposed control scheme does not require an accurate mathematical model of system and the joint acceleration measurement. From the numerical simulation results, it shows that the proposed controller is feasible and all the outputs converge to small neighborhoods of the desired reference trajectories.

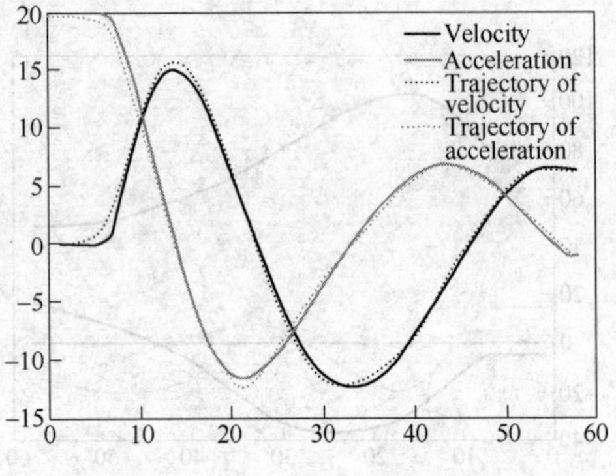

Fig. 5 Trajectory of velocity/acceleration for limb 2#

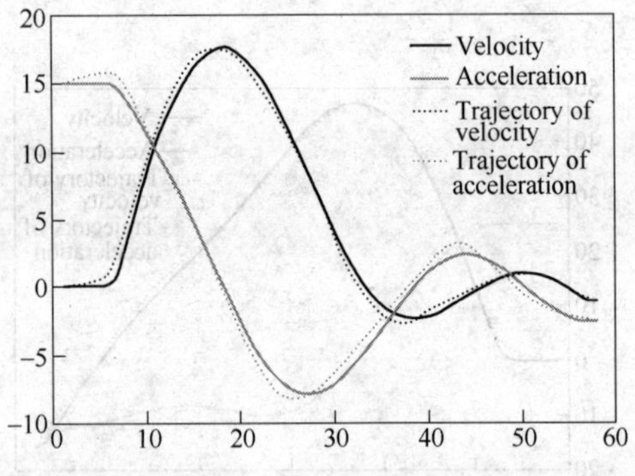

Fig. 6 Trajectory of velocity/acceleration for limb 3#

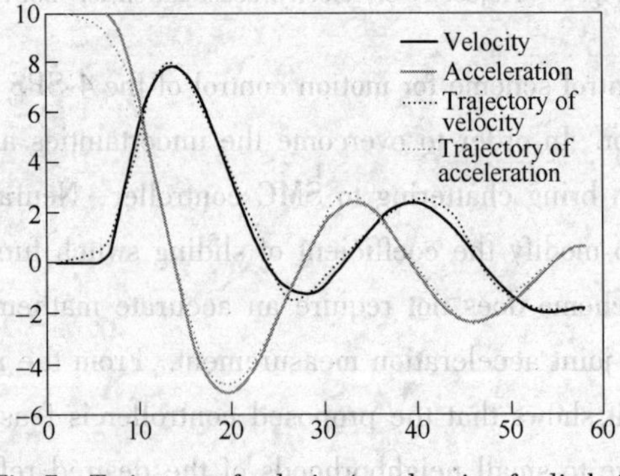

Fig. 7 Trajectory of velocity/acceleration for limb 4#

References

[1] D Stewart. A platform with six-degrees-of-freedom [J]. Proceedings of the Institute of Mechanical Engineering. London: UK. Vol. 180, No. 5, 1965: 371 ~ 386.

[2] Min kyn Park, Min Cheol Lee, Seok Jo Go. The design of sliding mode controller with perturbation observer for a 6-DoF parallel manipulator [C] //Proceedings of ISIE 2001. Pusan, KOREA, 2001: 1502 ~ 1508.

[3] Shijing Li, Zuren Feng, Hao Fang. Variable structure control for 6-6 parallel manipulators based on caseaded CMAC [C] //Proceedings of the 4th World Congress on Intelligent Control and Automation. Shanghai, P. R. China, 2002: 1939 ~ 1944.

[4] George K. I. Mann, W Brian. Surgenor. Model-free intelligent control of a 6-DoF Steward-Gough based parallel manipulator [C] //Proceedings of the 2002 IEEE International Conference on Control Applications, 18 ~ 20 September, Glasgow, Scotland, UK, 2002: 495 ~ 501.

[5] Huapeng Wu, Heikki Handroos. Hybrid fuzzy self-tunning PID controller for a parallel manipulator [C] //Proceedings of the 5th World Congress on Intelligent Control and Automation, 15 ~ 19 June, Hangzhou, P. R. China, 2004: 2545 ~ 2550.

[6] Ouyang P R, Zhang W J, Wu F X. Nonlinear PD control for trajectory tracking with consideration of the design for control methodology [C] //. Proceedings of the 2002 IEEE International Conference on Robotics & Automation, Washington D. C. , 2002: 30 ~ 36.

[7] Yiu Y K, Li Z X. PID and adaptive robust control of a 2-DoF over-actuated parallel manipulator for tracking different trajectory [C] //Proceedings IEEE International Symposium on Computation Intelligence in Robotics and Automation, Kobe, Japan, 16 ~ 20 July, 2003: 1052 ~ 1058.

[8] Su Y X, Dong Sun. Nonlinear PD synchronized control of parallel manipulators [C] //Proceedings of the 2005 IEEE International Conference on Robotics and Automation. Barcelona, Spain, 2005: 1374 ~ 1380.

[9] Yuefa Fang, Lung-Wen Tsai. Structure synthesis of a class of 4-DoF and 5-DoF parallel manipulators with identical limb structures [J]. The International Journal of Robotics Research, Vol. 21, No. 9, 2002: 799 ~ 810.

[10] ZHU Dachang, FANG Yuefa. Displacement analysis and synthesis of parallel manipulator via screw theory [J]. Robot, Vol. 27, No. 6, 2005: 539 ~ 544.

[11] ZHU Dachang, FANG Yuefa. Analysis identification of limb structure at special displacement for parallel manipulators [C] //Proceedings of the International Conference on Mechanical Engineering and Mechanics 2005. Nanjing, China. Science Press, USA. , 2005: 364 ~ 369.

[12] C L. Hwang. Fourier series neural-network-based adaptive variable structure control for servo-

systems with friction [J]. Proceedings of Industry Electronical and Engineering. Vol. 144, No. 6, 1997: 559 ~ 565.

[13] P Kachroo, M Tomizuka. Chattering reduction and error convergence in the sliding-mode control of a class of nonlinear systems [J]. IEEE Transaction of Automation Control. Vol. 32, No. 7, 1996: 1063 ~ 1068.

[14] Patra J C, Kot A C. Nonlinear dynamic system identification using chebyshev functional link artificial neural networks [J]. IEEE Trans. Systems, Man & Cybernetics. Vol. 32, No. 4, 2002: 505 ~ 511.

2.5 Robust Tracking Control of 4-SPS (PS) Type Parallel Manipulator Via Adaptive Fuzzy Logic Approach

The dynamic behavior of electro-hydraulic driven parallel manipulators is highly nonlinear system, the nonlinear behavior arising from load friction as well as the valve flow-pressure drop relationship. This paper is concerned with the robust tracking control of electro-hydraulic driven parallel manipulators with the model uncertainties. An adaptive fuzzy controller is used to estimate the uncertainties of electro-hydraulic system, including the payload variation and stiffness etc. Adaptive fuzzy logic and adaptive backstepping method are employed to provide the solution to the control problem. Simulation results from a electro-hydraulic driven 4-SPS (PS) type parallel manipulator demonstrate its strong robustness against a large of parameters variations and load disturbance and its capability of the trajectories tracking performance.

2.5.1 Introduction

Electro-hydraulic servo system is widely used in many industrial applications because of their high power-to-weight ratio, high stiffness, and high payload capability. However, the dynamic behavior of these systems is highly nonlinear due to phenomena such as nonlinear servo valve flow-pressure characteristics, variations in trapped fluid volumes and associate stiffness, and servo valve characteristics near null. To overcome the drawbacks, many different control techniques have been used to control position or force for a hydraulic actuator driven by a servo valve, including traditional PID controllers[1,2], recursive Lyapunov design method[3], controllers based on adaptive neural networks[4], and controllers based on quantitative feedback theory[5]. On the other hand, parallel manipulators usually adopt electro-hydraulic servo mechanisms as actuators which have high payload and precision at the same time.

In order to account for the nonlinear characteristic with structure uncertainties which are referred to as parametric uncertainties, the adaptive control strategy via feedback linearization can be used, which has undergone rapid developments in the past decade[6~8]. As for unstructured uncertainties, which are from modeling errors and external disturbances, deterministic robust control method can be used[9,10], if there is prior knowledge of the bound on the unstructured uncertainties. But the load acting on platform of parallel manipulators is uncertainties, then the adaptive control method and the deterministic robust control method can not be used to design controller for those systems. Several stable adaptive neural network control approaches are developed[11,12] and Lyapunov's stability theory was applied in designing adaptive neural network controller[13]. From a mathematical control perspective, fuzzy logic systems can be also used in a similar setting with neural networks, even combine with these two methods[14~16]. The fuzzy logic systems are used to uniformly approximate the unstructured uncertain functions in the designed process by use of the universal approximation properties of the certain classes of fuzzy systems, which were proposed in[17,18]. Recently, several stable robust adaptive neural network and fuzzy controllers have been studied for the systems with unstructured uncertainties[19,20]. However, the electro-hydraulic servo system has a lot of parameters which needed to be tuned in the online learning laws. For a fuzzy logic systems, there have many rule bases are used which are used to approximate the uncertain nonlinear functions of electro-hydraulic servo system. Obviously, this process with higher order system and time-consuming is unavoidable during work.

In this paper, we present an adaptive fuzzy controller to estimate the uncertainties of electro-hydraulic servo system and adaptive backstepping method is employed to provide the solution to the control problem with robust control strategy. This paper is organized as follows. In section 2, we will present 4-SPS (PS) type parallel manipulator single-channel electro-hydraulic servo model. Adaptive fuzzy logic control is proposed in section 3. In section 4, robust adaptive fuzzy logic controller is developed. Simulations are given to demonstrate the effectiveness of schemes. The final section contains conclusion.

2.5.2 Model of 4-SPS (PS) Type Parallel Manipulator

2.5.2.1 Single channel eletro-hydraulic servo model

The mathematical model of an asymmetric servo-valve system for the single channel hydraulic cylinder shown in Fig. 1 is as follow. Without any elastic load[21].

$$Y(s)=\frac{(K_{x\alpha}/A_{me})x-\dfrac{K_{t\alpha}}{A_eA_{me}}\left(\dfrac{V_e}{4\beta_eK_{t\alpha}}s+1\right)f_e}{(s^2/\omega_{he}^2+2\xi_{he}/\omega_{he}+1)s} \tag{1}$$

where $K_{x\alpha}$ is the flow gain, m^2/s; A_{me} is the average area of piston, m^2; $K_{t\alpha}$ is the total flow-pressure coefficient, $m^2/(s\cdot N)$; A_e is the equivalent area of hydraulic cylinder, m^2; β_e is the modulus of elasticity, MPa; f_e is the equivalent outer interference force, N; x is the open size of valve, m; ω_{he} is the inherent frequency of hydraulic servo system; ξ_{he} is the damp ratio.

$$\omega_{he}=2\sqrt{\frac{A_eA_{me}\beta_e}{V_em}}\quad(\mathrm{rad/s})$$

$$\xi_{he}=K_{t\alpha}\sqrt{\frac{m\beta_e}{V_eA_eA_{me}}}+\frac{B_p}{4}\sqrt{\frac{V_e}{m\beta_eA_eA_{me}}}$$

where m is the loading mass, kg; B_p is the viscidity damping coefficient of load, N · s.

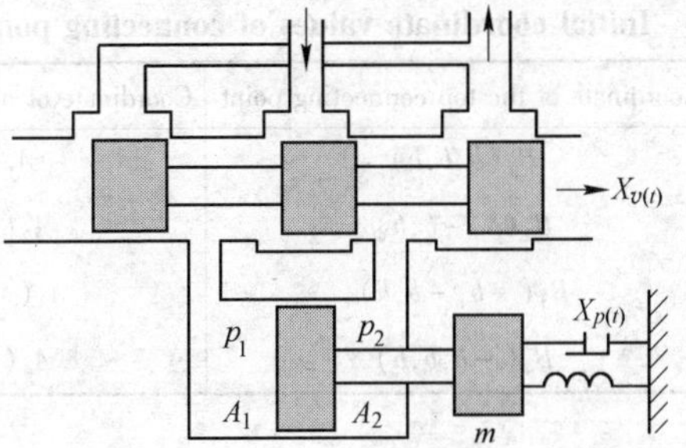

Fig. 1 Model of single channel asymmetric servo valve

2. 5. 2. 2 Model of 4-SPS (PS) type parallel manipulator

A schematic representation of the 4-SPS (PS) type parallel manipulator is shown in Fig. 2, where the fixed platform is labeled as FP and the moving platform is MP. Indeed MP is connected to FP through four identical leg mechanisms, and is driven by the corresponding hydraulic servo actuators. An especial leg which only provides restricts to MP and connected MP with FP in the middle of each platform. A prismatic joint connected MP by a spherical joint and the other side is fixed on FP.

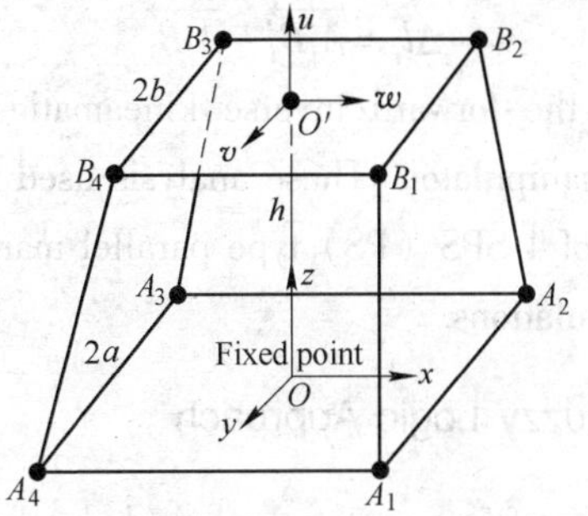

Fig. 2 Model of 4-SPS (PS) type parallel manipulator

The parameters are shown in Fig. 2. The initial coordinate values of connecting point with fixed platform and moving platform of limbs are listed in Table 1.

Table 1 Initial coordinate values of connecting point limbs

Number of limbs	Coordinate of the top connecting point	Coordinate of nether connecting point
1	$B_1(b,b,h)$	$A_1(a,a,0)$
2	$B_2(b,-b,h)$	$A_2(a,-a,0)$
3	$B_3(-b,-b,h)$	$A_3(-a,-a,0)$
4	$B_4(-b,b,h)$	$A_4(-a,a,0)$

Assume that the rotational angle of three dimensions denoted as α, β and γ respectively, and the translational along with z is d, then the transform coordinate of B_i can be gotten as follows.

$$P_{B_i'} = R_{B_o} P_{B_i} + \text{Trans}(z,d) P_{B_i} \tag{2}$$

where $R_{B_o} = R(z,\gamma)R(y,\beta)R(x,\alpha) = \begin{bmatrix} c\gamma & -s\gamma & 0 \\ s\gamma & c\gamma & 0 \\ 0 & 0 & 1 \end{bmatrix} \begin{bmatrix} c\beta & 0 & s\beta \\ 0 & 1 & 0 \\ -s\beta & 0 & c\beta \end{bmatrix}$.

$$\begin{bmatrix} 1 & 0 & 0 \\ 0 & c\alpha & -s\alpha \\ 0 & -s\alpha & c\alpha \end{bmatrix}$$

Let the extend displacement of each actuators are $\Delta l_i (i = 1 \sim 4)$

$$\Delta l_i = \overrightarrow{A_i B_i'} - \overrightarrow{A_i B_i} \tag{3}$$

Eq. (3) denotes the forward/inverse kineamatic solution with 4-SPS (PS) type parallel manipulator. These analysis used in this paper to build a mathematic model of 4-SPS (PS) type parallel manipulator by Matlab@ SimMechanics in simulations.

2.5.3 Adaptive Fuzzy Logic Approach

A fuzzy system consists of four parts: the knowledge base, the fuzzifier, the fuzzy inference engine working on fuzzy rules and the defuzzifier. The knowledge base for the fuzzy logic system comprises a collection of fuzzy If-then rules. The four channels electro-hydraulic servo system of 4-SPS (PS) type parallel manipulator if-then rules are of the following form

$$R = U_{l=1}^{4} R_l,\ R_l: \text{if} \quad x_1 \text{ is } A_1^l \quad x_2 \text{ is } A_2^l \quad \cdots \quad x_q \text{ is } A_q^l,$$

$$\text{then } y_1 \text{ is } B_1^l \quad y_2 \text{ is } B_2^l \quad \cdots \quad y_p \text{ is } B_p^l \tag{4}$$

where R is the union of fuzzy rules in which each rule is of the form of R_l, $\boldsymbol{x} = [x_1, x_2, \cdots, x_q]^{\mathrm{T}}$ and $\boldsymbol{y} = [y_1, y_2, \cdots, y_p]^{\mathrm{T}}$ are the input and output vectors of the fuzzy system respectively and $p = q = 4$. A_i^l and B_j^l are the linguistic variables of the fuzzy set in the subspace U_i and V_j, described by their membership functions $\mu_{A_i^l}(x_i)$ and $\mu_{B_j^l}(y_j)$, $(i = 1, \cdots, q)$ and $(j = 1, \cdots, q)$.

$$U = U_1 \times U_2 \times \cdots U_q, U_i \in \mathbf{R}^q, \ V = V_1 \times V_2 \times \cdots V_p, V_j \in \mathbf{R}^p \tag{5}$$

The fuzzifier maps x in the input space $U \in \mathbf{R}^q$ to a fuzzy set A^l in the output space $V \in \mathbf{R}^p$. The fuzzy inference engine performs a mapping from fuzzy sets in V to fuzzy sets in U, based on fuzzy if-then rules in fuzzy rule base and the compositional rule of inference.

The output of fuzzy logic system with singleton fuzzifier is of the following type.

$$y_j = \frac{\sum_{l=1}^{p} \prod_{i=1}^{q} \mu_{A_i^l}(x_i) y_j^l}{\sum_{l=1}^{M} \prod_{i=1}^{q} \mu_{A_i^l}(x_i)} \tag{6}$$

where y_j^l is a point on V_j at which $\mu_{B_j^l}(y_j)$ achieves its maximum value which is equal to 1.

2.5.3.1 Dynamic of 4-SPS (PS) type parallel manipulator

The linear parameterization of manipulator dynamics can be obtained as follows.

$$\boldsymbol{M}(\boldsymbol{q})\ddot{\boldsymbol{q}} + \boldsymbol{C}(\boldsymbol{q}, \dot{\boldsymbol{q}})\dot{\boldsymbol{q}} + \boldsymbol{G}(\boldsymbol{q}) = \boldsymbol{\tau} \tag{7}$$

where $\boldsymbol{q}$ is the vectors of generalized joint position, $\boldsymbol{q} \in \mathbf{R}^4$; $\boldsymbol{M}(\boldsymbol{q})$ is the inertia matrix; $\boldsymbol{C}(\boldsymbol{q}, \dot{\boldsymbol{q}})$ is the Coriolis and centrifugal effects; $\boldsymbol{G}(\boldsymbol{q})$ is the gravitational term; $\boldsymbol{\tau}$ represents the vector of generalized control input forces.

The matrices in Eq. (6) satisfy the following properties.

Property 1 $\boldsymbol{M}(\boldsymbol{q})$ is a symmetric positive definite matrix.

Property 2 $(\dot{\boldsymbol{M}} - 2\boldsymbol{C})$ is a skew-symmetric matrix.

Property 3 $C(x,y)z = C(x,z)y \quad \forall x,y,z \in \mathbf{R}^{n\times 1}$.

Property 4 $\exists C_M$ s. t. $\|C(x,y)\| \leqslant C_M\|y\| \quad \forall x,y \in \mathbf{R}^{n\times 1}$.

For the manipulator dynamic model given by Eq. (6), the following assumptions are made.

Assumption 1 The maximum allowable torques for each joint $\tau_i^{\max}$ for $i =$ 1, 2, 3, 4 are known.

Assumption 2 The desired joint trajectory $q_{di}(t)$ is bounded.

The control objective can be stated as follow: given desired trajectories $q_d(t) = [q_{d1}, \cdots, q_{d4}]^T$, determine a control law τ_{eq} which is a function of position only achieves the following two goals in the presence of uncertainties and actuator constrains.

$$\lim_{t\to\infty} q(t) = q_d, \ |\tau_i(t)| \leqslant \tau_i^{\max}, \forall t \geqslant 0 \quad i = 1, \cdots, 4 \tag{8}$$

2.5.3.2 Robust adaptive fuzzy controller design

Assuming the desired trajectory for i^{th} limb to be twice differentiable, define the following relations.

$$\tilde{q}_i = q_i - q_{di} \tag{9}$$

$$\dot{q}_{ri} = \dot{q}_{di} - \Lambda_i \tilde{q}_i \tag{10}$$

$$s_i = \dot{q}_i - \dot{q}_{ri} = \dot{\tilde{q}}_i + \Lambda_i \tilde{q}_i \tag{11}$$

where $\tilde{q}_i$ is the tracking error; $\dot{q}_{ri}$ is the reference actuator velocity and s_i is the residual error for the ith limb system. Substituting (9) ~ (11) into the dynamic equation, described by (7), results in

$$M(q_i)\dot{s}_i = -C(q_i,\dot{q}_i)s_i - \Lambda(q_i,\dot{q}_i) + \tau_i - (M(q_i))\ddot{q}_{ri} + C(q_i,\dot{q}_i)\dot{q}_{ri} + G(q_i) \tag{12}$$

According to the general approximation property of the adaptive fuzzy approach, we can write

$$M(q_i)\ddot{q}_{ri} + C(q_i,\dot{q}_i)\dot{q}_{ri} + G(q_i) = \psi_i^* \varphi_i(q_i,\dot{q}_i,\dot{q}_{ri},\ddot{q}_{ri}) + w_i \tag{13}$$

where w_i is the minimum approximation errors of the adaptive fuzzy which can be made sufficienty small by choosing the approximation fuzzy basis functions $\boldsymbol{\varphi}(\cdot)$ which defined as following

$$\boldsymbol{\varphi}(x) = \frac{\sum_{l=1}^{p} \prod_{i=1}^{q} \mu_{A_i^l}(x_i)}{\sum_{l=1}^{M} \prod_{i=1}^{q} \mu_{A_i^l}(x_i)} \tag{14}$$

Then Eq. (6) can be written as

$$y(x) = \boldsymbol{\theta}^{\mathrm{T}} \boldsymbol{\varphi}(x) \tag{15}$$

where $\boldsymbol{\theta} = (y_j^1, \cdots, y_j^l)^{\mathrm{T}}$ is the parameter vector, and $\boldsymbol{\varphi}(x) = (\varphi^1(x), \cdots, \varphi^l(x))^{\mathrm{T}}$ is called as basis function.

The control law could be approximated by fuzzy logic system of the represented in (15).

$$u = \boldsymbol{\theta}^{\mathrm{T}} \boldsymbol{\varphi}(x) \tag{16}$$

The elements of the parameter vector of the controller θ is varied based on some adaptation law to cope with uncertain plant parameter variation. As the fuzzy technique takes the form of linear parameter model that is required by the classical adaptive control, so let the adaptation law is given by

$$\dot{\boldsymbol{\theta}} = -\gamma \boldsymbol{\varphi}(x) \boldsymbol{B}^{\mathrm{T}} P e \tag{17}$$

where γ is the adaptation rate, P is the solution of the following Riccati-like equation

$$PA + A^{\mathrm{T}}P + Q - \frac{2}{r} P\boldsymbol{B}\boldsymbol{B}^{\mathrm{T}}P + \frac{1}{\rho^2} P\boldsymbol{B}\boldsymbol{B}^{\mathrm{T}}P = 0 \tag{18}$$

where r is a positive weighting factor and ρ is an attenuation level. $\boldsymbol{A}$ and $\boldsymbol{B}$ matrices are given as

$$\boldsymbol{A} = \begin{bmatrix} 0 & 1 & 0 & \cdots & 0 \\ 0 & 0 & 1 & \cdots & 0 \\ 0 & 0 & 0 & \cdots & 0 \\ \vdots & \vdots & \vdots & & \vdots \\ -k_n & -k_{n-1} & -k_{n-2} & \cdots & -k_1 \end{bmatrix} \quad \boldsymbol{B} = \begin{bmatrix} 0 \\ 0 \\ 0 \\ \vdots \\ 1 \end{bmatrix}$$

Due to the approximation error caused by the fuzzy system, modification of the adaptation law is necessary so that the system does not become unstable with large approximation error. The (17) is modified by adaptation law with continuous switching leakage

$$\dot{\boldsymbol{\theta}} = -\gamma\boldsymbol{\varphi}(x)\boldsymbol{B}^{\mathrm{T}}Pe - \gamma\sigma_{\mathrm{s}}\boldsymbol{\theta} \tag{19}$$

where σ_{s} is called continuous switching function and represented as

$$\sigma_{\mathrm{s}} = \begin{cases} 0 & \text{if } |\boldsymbol{\theta}(t)| < M_0 \\ \sigma_0\left(\dfrac{|\theta(t)| - M_0}{|\theta(t)|}\right) & \text{if } M_0 \leqslant |\theta(t)| \leqslant 2M_0 \\ \sigma_0 & \text{if } |\theta(t)| > 2M_0 \end{cases} \tag{20}$$

where σ_0 and M_0 are design constants.

Adaptive fuzzy system shown as Fig. 3.

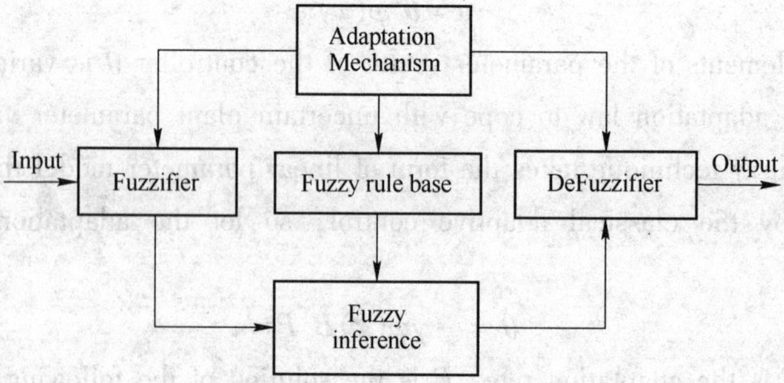

Fig. 3 Adaptive fuzzy system

2.5.4 Experimental Results

Consider the 4-SPS (PS) type parallel manipulator which parameters are given as following.

By using Matlab software, the reference model is build shown as Fig. 4 and the actual simulation model is build shown as Fig. 5. The input signals of four actuators are given as.

From Fig. 6 to Fig. 9 it is evident that the proposed controller is quite effective for elimination much of the flattering time delay caused by the hydraulic system.

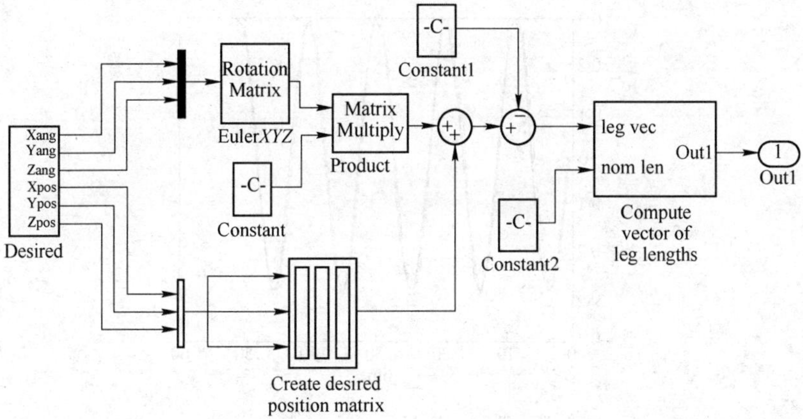

Fig. 4 Reference model of 4-SPS (PS) type parallel manipulator

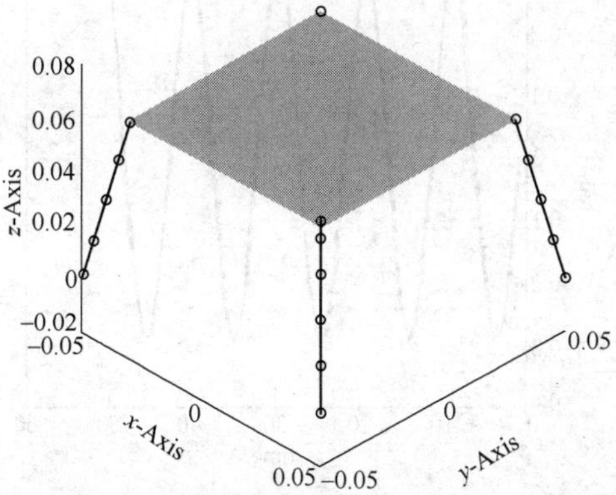

Fig. 5 Actual simulation model of 4-SPS (PS) type parallel manipulator

2.5.5 Conclusions

In this investigation we have proposed a robust adaptive fuzzy controller for a class of nonlinear systems like parallel manipulators driven by hydraulic actuators. The proposed controller has be designed for hydraulically actua-

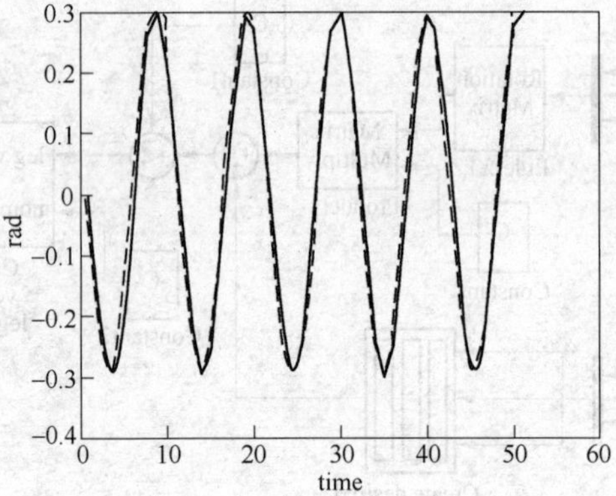

Fig. 6 Trace tracking of the 1th actuator

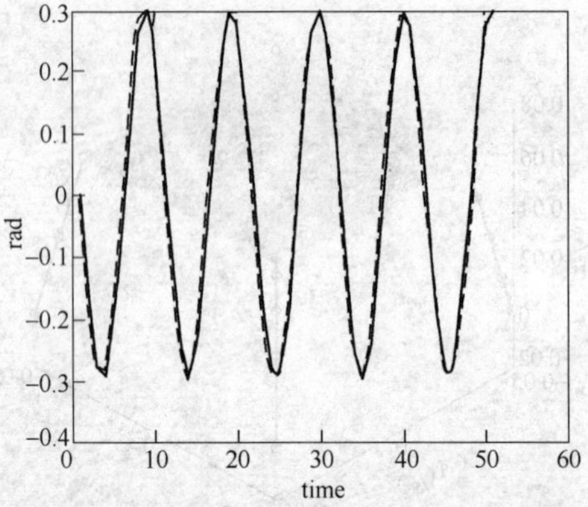

Fig. 7 Trace tracking of the 2th actuator

ted 4-SPS (PS) type parallel manipulator which has four channels hydraulic servo system. The experimental results exhibit that the robust adaptive controller performs quite good tracking of the linear actuators compared to the perfect trace and the proposed control scheme results a very stable locomotion as shown in the trace tracking of moving platform of 4-SPS (PS) type parallel manipulator.

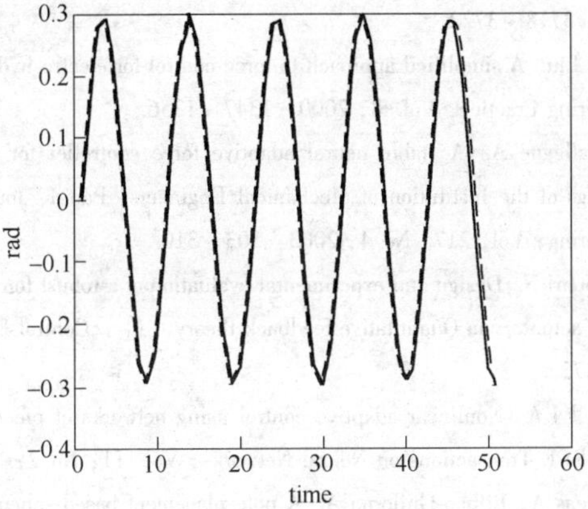

Fig. 8 Trace tracking of the 3th actuator

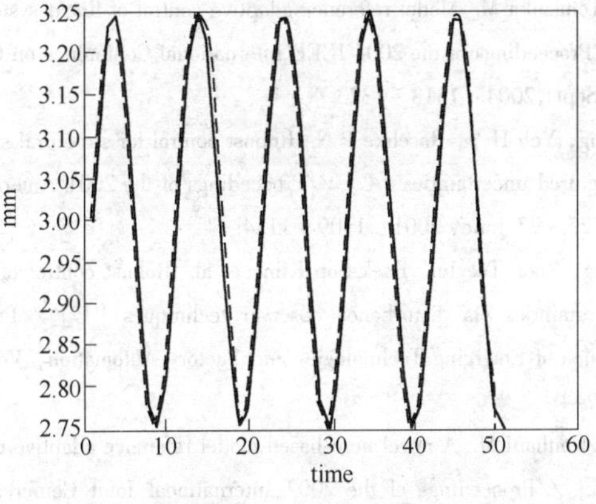

Fig. 9 Trace tracking of the 4th actuator

References

[1] Alleyne A, Rui Liu, Wright H. On the limitations of force tracking control for hydraulic active suspensions [C] //Proceedings of the 1998 American Control Conference, Philadelphia, PA, USA, Vol. 1, 1998: 43 ~ 47.

[2] Niksefat N, Sepechri N. Robust force controller design for a hydraulic actuator based on exper-

imental input-output data [C] //Proceedings of the American Control Conference, San Diego, CA, 1999: 3718 ~ 3722.

[3] Alleyne A, Rui Liu. A simplified approach to force control for electro-hydraulic systems [J]. Control Engineering Practice, Vol. 8, 2000: 1347 ~ 1356.

[4] Daachi B, Benallegue A. A stable neural adaptive force controller for a hydraulic actuator [J]. Proceedings of the Institution of Mechanical Engineers, Part I: Journal of Systems and Control Engineering, Vol, 217. No. 4, 2003: 303 ~ 310.

[5] Niksefat N, Sepehri N. Design and experimental evaluation of a robust force controller for an electro-hydraulic actuator via Quantitative feedback theory [J] . Control Engineering Practice, 2000: 3718 ~ 3722.

[6] Choi J Y, Farrell J A. Nonlinear adaptive control using networks of piecewise linear approximators [J] . IEEE Transactions on Neural Networks, Vol. 11, No. 2 , 2000: 390 ~ 401.

[7] M de la Sen, lbeas A, Bilbao-Guilerna A. A pole-placement based scheme for robustly stable adaptive control of continuous linear systems with multiestimation [C] //Proceedings of the 2005 IEEE International Symposium on Intelligent Control, Mediterrean Conference on Control and Automation, 27 ~ 29 June, 2005: 1055 ~ 1061.

[8] Kanamori M, Tomizuka M. Model reference adaptive control of linear systems with input saturation [C] //Proceedings of the 2004 IEEE International Conference on Control Application, Vol. 2, 2 ~ 4 Sept, 2004: 1318 ~ 1323.

[9] Sheng-Guo Wang, Yeh H Y, Roschke P N. Robust control for structural systems with parametric and unstructured uncertainties [C] //Proceedings of the 2001 American Control Conference, Vol. 2, 25 ~ 27 June, 2001: 1109 ~ 1114.

[10] Seong-Ho Song, Yoon-Tae Im, Baek-Sop Kim, et al. Robust control of linear systems with nonlinear uncertainties via disturbance observer techniques [C] //Proceedings ETFA'03 IEEE Conference of Emerging Technologies and Factory Automation, Vol. 1, 16 ~ 19 Sept, 2003: 241 ~ 244.

[11] Menhaj M B, Rouhani M. A novel nero-based model reference adaptive control for a two link robot arm [C] //Proceedings of the 2002 International Joint Conference on Neural Networks, Vol. 1, 12 ~ 17 May, 2002: 47 ~ 52.

[12] Martins N A, Figueiredo M F, Goncalves P C, et al. Trajectory tracking performance in task space of robot manipulators: an adaptive neural controller design [C] //2005 IEEE/RSJ International Conference on Intelligent Robots and Systems, 2 ~ 6 Aug, 2005: 1241 ~ 1246.

[13] Meng Joo Er, Yang Gao. Robust adaptive control of robot manipulators using generalized fuzzy neural networks [J] . IEEE Transactions on Industrial Electronics, Vol. 50, No. 3, 2003: 620 ~ 628.

[14] Shibli M. Direct adaptive control for underactuated mechatronic systems using fuzzy systems and neural networks: a pendubot case. Canadian Conference on Electrical and Computer En-

gineering, 2006: 1490 ~ 1493.

[15] Changhong Wang, Baoming Feng, Guangcheng Ma, et al. Robust tracking control of space robots using fuzzy neural network [C] //Proceedings of 2005 IEEE International Symposium on Computational Intelligence in Robotics and Automation, 27 ~ 30 June, 2005: 181 ~ 185.

[16] Shu-Huan Wen, Qi-Guang Zhu. Generalized dynamic fuzzy neural network-based tracking control of robot manipulators [C] //Proceedings of 2004 International Conference on Machine Learning and Cybernetics, Vol. 2, 26 ~ 29 Aug, 2004: 812 ~ 816.

[17] Wang L X. Fuzzy systems are universal approximators [C] //IEEE Proceedings of International Conference on Fuzzy Systems, San Diego, CA, 1992: 1163 ~ 1170.

[18] Wang L X, Mendel J M. Fuzzy basis functions universal approximation, and orthogonal least-squares learning [J]. IEEE Transactions on Neural Networks, Vol. 3, 1992: 807 ~ 814.

[19] Yansheng Yang, Gang Feng, Junsheng Ren. A combined backstepping and small-gain approach to robust adaptive fuzzy control for strictfeedback nonlinear systems [J]. IEEE Transactions on Systems, Man and Cybernetics, Vol. 34, No. 3, 2004: 406 ~ 420.

[20] Kuang-Yow Lian, Chian-Song Chiu, Liu P. Semi-decetralized adaptive fuzzy control for cooperative multirobot systems with H'motion/internal force tracking performance [J]. IEEE Transaction on Systems, Man and Cybernetics, Part B, Vol. 32, No. 3, 2002: 269 ~ 280.

[21] ZHU Dachang. Sliding mode synchronous control for fixture clamps system driven by hydraulic servo systems [J]. Proc. IMechE Part C: Journal of Mechanical Engineering Science, Vol. 221, 2007: 1039 ~ 1046.

3　全柔顺并联机构空间构型综合与刚度研究

3.1　2RPU-2SPS 全柔顺并联机构构型设计及刚度研究

2RPU-2SPS 型并联机构是由 RPU 支链和 SPS 支链组成的非对称结构，是具有 2 转动 2 平移运动特性的并联机构。本文基于 2RPU-2SPS 并联机构的运动特性，采用替换法设计出与之对应的全柔顺并联机构。应用 Ansys 软件建立 2RPU-2SPS 型柔性并联机构、全柔顺并联机构的支链模型，并对全柔顺并联机构的支链进行拓扑优化，进而设计出具有全柔顺特性的新型的 2RPU-2SPS 型全柔顺并联机构。实验仿真对比结果表明：全柔顺支链在整体刚度性能方面优于柔性并联机构支链，该设计方法为全柔顺并联机构整体设计提供了理论依据。

3.1.1　引言

并联机构由于刚度大、精度高、承载能力强、操作速度高等优点，得到迅猛的发展和广泛的应用。但是，传统的并联机构由于其运动支链是由刚性连杆以运动副连接而成的，这在高速、高精度、微型化等要求下不可避免地会暴露其固有缺陷，比如自身惯性所带来的振动，运动副与刚性连杆连接所产生的间隙、摩擦、润滑、装配、误差等。柔顺机构的出现为这些问题的解决提供了一种新的方法。柔顺机构主要依靠机构中柔性构件的自身变形来实现机构的运动、力和能量的传递和转换，从而实现机构全部的运动和功能。因此，柔顺机构大大减少或消除了上述问题。目前，国内外许多学者将柔性机构相关理论与并联机构理论相结合，对空间柔顺并联机构进行了深入的研究。Jae-Jong Lee 等设计了六自由度自适应定位工作台[1]和五自由度自适应承片台[2]。王华等[3]设计出一种整体式空间三自由度柔顺并联微动平台。荣伟彬等[4]提出了一种由压电陶瓷驱动的具有 3-PPSR 结构的六自由度柔顺并联机构。目前，对空间柔顺并联机构的大部分研究都只是将并联机构各运动支链的运动副以柔性铰链代替，再用柔性连杆将各个柔性铰链连接在一起形成相对应的柔性支链，从而构成柔顺并联机构，如杨启志等[5]所设计的一种非对称全柔性 3-RRRP 平移微动并联机构，通过这种方法设计的柔顺并联机构将并联机构与柔顺机构理论相结合，可实现所要求的运动和功能，但在超精密定位等要求高精度的应用领域，还存在些许不足。因此，设计出一种具有更高刚度的柔顺并联机构类型，就显得很有必要。

文中对 2RPU-2SPS 并联机构的运动特性进行了分析，同时根据 2RPU-2SPS 并联机构[6]，采用替换法设计出 2RPU-2SPS 柔性并联机构，并在此基础上设计出 2RPU-2SPS 全柔性并联机构。利用 Ansys 软件分析柔性机构和全柔顺并联机构的支链受力情况，同时采用 Ansys 软件的拓扑优化功能，使得所设计的全柔顺并联机构的支链结构更加合理。通过优化分析对比，得出所设计的全柔顺并联支链在刚度上有较大提高，为超精密定位平台的精度研究提供可参考的理论依据。

3.1.2 结构简介

2RPU-2SPS 是非对称的四自由度并联机构构型形式，它是由 2 条 RPU 支链和 2 条 SPS 支链构成的，如图 1 所示，并且相同的支链相邻地组装在基座上。RPS 支链由转动副（R）、移动副（P）和虎克铰（U）组成，且具有 $^{w}\mathrm{R} \perp {}^{v}\mathrm{P} \perp {}^{wv}\mathrm{U}$ 的几何关系，其中 U 副是由两个转动副（即 $^{w}\mathrm{R}$ 和 $^{v}\mathrm{R}$）组成。SPS 支链是由两个球面副（S）和移动副（P）组成。S 是由 3 个 R 副（即 $^{w}\mathrm{R}$, $^{v}\mathrm{R}$, $^{w}\mathrm{R}$）组成。

3.1.3 运动特性分析

在分析并联机器人机构中，螺旋理论是一个非常有效的工具[7,8]。螺旋的基本形式可以表示为：$\hat{\boldsymbol{\$}} = [\boldsymbol{s} \quad \boldsymbol{s}_0]^{\mathrm{T}}$，其中，$\boldsymbol{s}$ 为沿螺旋轴线方向的单位矢量，$\boldsymbol{s}_0$ 为对偶数，当螺旋 $\boldsymbol{\$}$ 的两矢量表示为标量时，可用 *plcüker* 坐标（l, m, n, o, p, q）来表示。如果有两个螺旋 $\boldsymbol{\$}$ 和 $\boldsymbol{\$}_r$，若满足下述条件：$\boldsymbol{\$} \circ \boldsymbol{\$}_r = 0$，则 $\boldsymbol{\$}$ 和 $\boldsymbol{\$}_r$ 互为反螺旋，式中“∘”表示互易积。从物理意义上讲，互易积为零的两个螺旋，一个可用来表示物体的运动，另一个则表示物体受力情况。零节距的力

螺旋描述为力，无穷大节距的力螺旋表示为力偶。

在动平台上建立基坐标系 $B-xyz$，如图 2 所示，x 轴和 y 轴在动平台内，z 轴垂直于动平台。在定平台上建立支链 RPU 的坐标系$B_1-x_1y_1z_1$，其中，z_1 垂直于定平台，x_1 轴与转动副 R 的轴线平行。同时建立支链 SPS 的坐标系 $B_2-x_2y_2z_2$，它的 x、y 和 z 轴分别和基座上的组成 S 副的 3 个 R 副的轴线同轴，即 x 轴与 R_1 同轴，y 轴与 R_2 同轴，z 轴与 R_3 同轴。

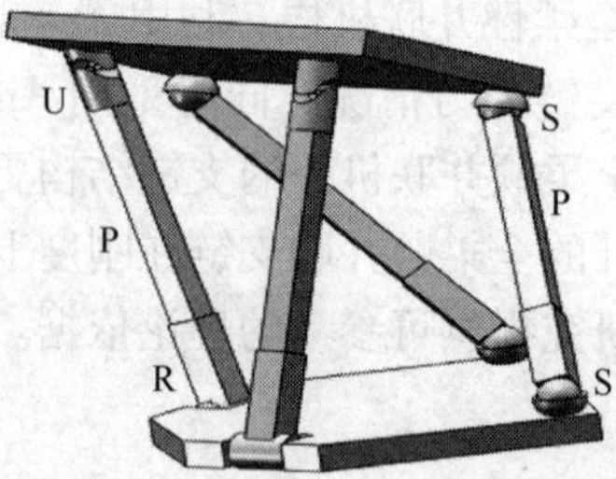

图 1　2RPU-2SPS 型并联机构构型

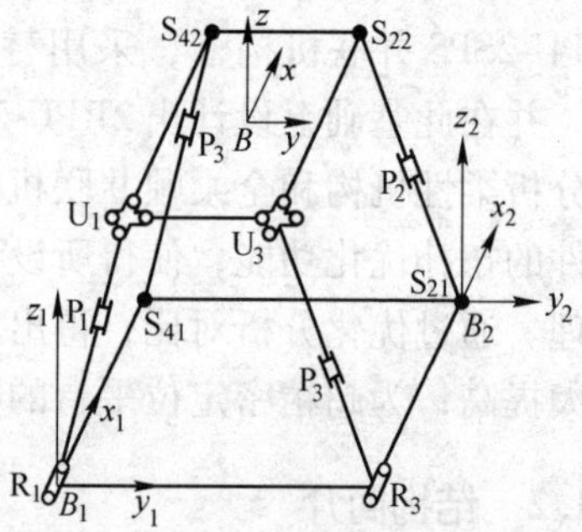

图 2　2RPU-2SPS 型并联机构结构简图与坐标系建立

在 RPU 支链上，U 是由两个 R 副组成，所以支链的运动螺旋系为：

$$\$_{11}=(1,0,0,0,0,0),\ \$_{12}=(0,0,0,0,0,1),$$

$$\$_{13}=(1,0,z,0,\alpha_3,0),\ \$_{14}=(0,0,1,0,\alpha_4,\beta_4)$$

该支链的末端杆有 4 个自由度，并受到两个约束。对应这两个约束的反螺旋可以由螺旋的互易积求零求得，其运动反螺旋为：

$$\$_r^1=(0,1,0,0,0,0),\ \$_r^2=(0,0,0,1,0,0)$$

由分析得出，该支链提供一个约束力偶和一个约束力，约束了支链的一个绕 y 轴转动和一个沿 x 轴移动的自由度。由于机构具有两个相同的 RPU 支链，所以均对动平台提供一个约束力偶和一个约束力，并且，所有的约束力偶相互平行和力线矢共轴。

在 SPS 支链上，S 由 3 个 R 副所组成，该支链的运动螺旋系为：

$$\$_{31}=(1,0,0,0,0,0),\ \$_{32}=(0,1,0,0,0,0),$$
$$\$_{33}=(0,0,1,0,0,0),\ \$_{34}=(0,0,0,0,0,1)$$

其运动反螺旋为0，则表明支链 SPS 对动平台不提供约束，为无约束支链，只具有支撑作用。

综上所述，在 4 条支链中，只有两条 RPU 支链对动平台提供 2 个约束，限制了动平台绕 y 轴的转动和沿 x 轴的移动。故 2RPU-2SPS 并联机构具有 4 个自由度[9]，即绕 x、z 轴转动和沿 y、z 轴的移动。

3.1.4　2RPU-2SPS 柔性并联机构的设计及支链刚度分析

3.1.4.1　柔性并联机构设计的原则

并联机构转化为柔性并联机构的原则如下：

(1) 组成柔性并联机构的各个柔性支链的相对位置与并联机构中的位置一致；

(2) 各柔性并联机构支链上的运动副与并联机构中的运动副相对应；

(3) 柔性并联机构上的各柔性运动副的位置设计满足并联机构所要求的几何条件。

3.1.4.2　2RPU-2SPS 柔性并联机构的设计

文中采用替换法在 2RPU-2SPS 并联机构的基础上设计了 2RPU-2SPS 柔性并联机构，满足并联机构转化为柔性并联机构设计原则。首先，设计相应的 RPU 和 SPS 型柔性支链，然后，相同的支链相邻地固定在定平台上，即形成 2RPU-2SPS 型柔性并联机构平台，如图 3 所示。

3.1.5　2RPU-2SPS 全柔顺并联机构设计

根据柔性并联机构设计全柔顺并联机构时，应注意以下两点：

(1) 主要几何尺寸设定一致，便于进行模态刚度的比较；

(2) 设定相同的约束及载荷。

在 2RPU-2SPS 型柔性并联机构的基础上设计与之相对应的 2RPU-2SPS 型全柔顺并联机构，首先设计相应的 RPU 和 SPS 型全柔

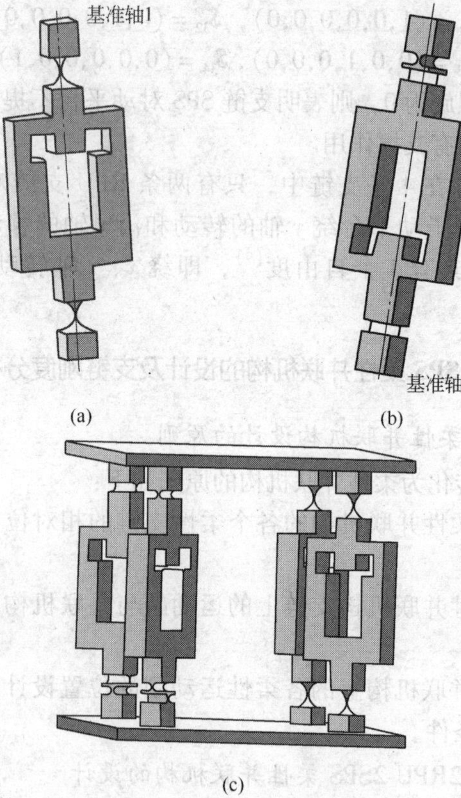

图 3 2RPU-2SPS 全柔性支链及其柔性并联机构构型

（a）SPS 柔性支链；（b）RPU 柔性支链；

（c）2RPU－2SPS 柔性并联机构

顺并联支链：全柔顺并联支链是在选定合适的一整块材料上通过先进切割技术分别切割出各个相应的柔性铰链和柔性连杆，如图 4 所示。在选定材料上通过先进切割技术按要求切割出一个柔性虎克铰 U、一个柔性移动副 P、一个柔性转动副 R，从而形成 RPU 型全柔顺并联支链，如图 4（a）所示；SPS 型柔顺支链是在选定材料上按要求切割出两个柔性球面副 S 和一个柔性移动副 P，如图 4（b）所示；按照 2RPU-2SPS 全柔顺并联机构的几何特性进行安装，可得 2RPU-2SPS 型全柔顺并联机构构型，如图 4（c）所示。

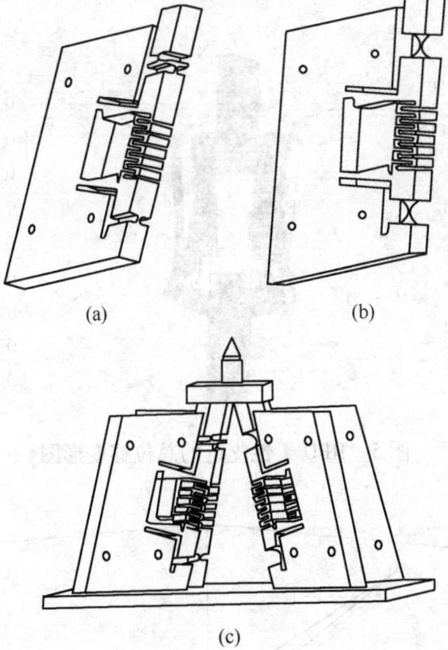

图4 2RPU-2SPS型全柔顺并联机构构型
(a) 初设计RPU全柔顺支链；(b) 初设计SPS全柔顺支链；
(c) 2RPU－2SPS全柔顺并联机构

3.1.6 2RPU-2SPS柔性及全柔顺并联机构刚度对比研究

3.1.6.1 RPU柔性支链和RPU型全柔顺支链的刚度分析及对比研究

应用Ansys软件从SolidWorks中导入RPU型柔性并联机构支链模型，定义机构材料属性和单元类型，进行网格划分。将支链上的一个支点设定自由度为0，并在支链上施加驱动力和外加载荷为1000N。

对机构的Ansys模型进行计算和后处理，得到机构支链总位移变形图如图5所示。

选取支链输出平面上的若干关键节点，在外力作用下这些节点的位移变形如图6所示。

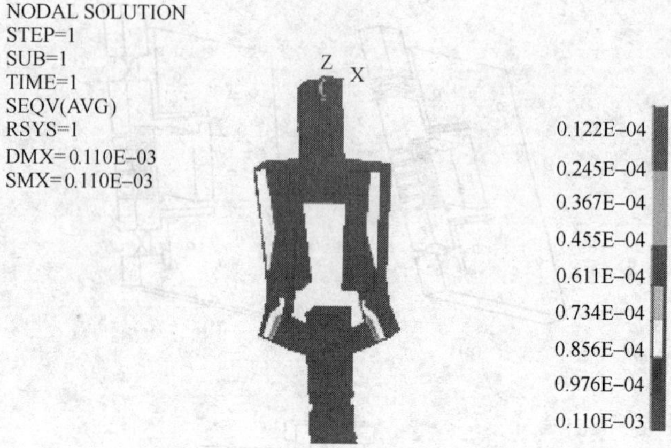

图 5 RPU 柔性支链的总位移变形图

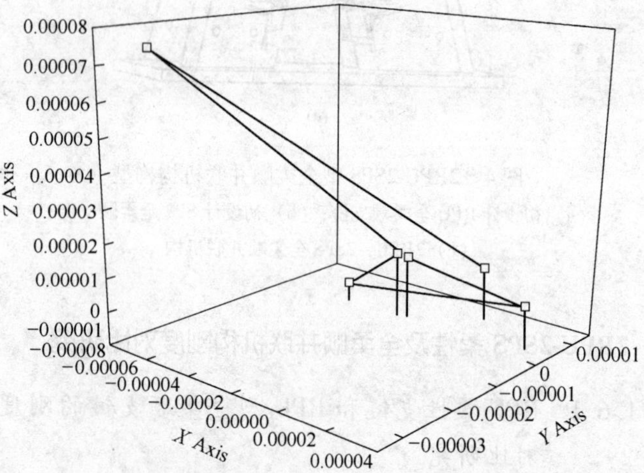

图 6 输出平面节点位移变形图

对图 6 中的 *UX*、*UY*、*UZ* 分别求平均值，可得到位移：$\overline{UX}=-0.23850E-04$，$\overline{UY}=0.14626E-04$，$\overline{UZ}=0.18428E-05$，由此可知支链输出平面的运动。

3.1.6.2 RPU 柔性支链的刚度计算

RPU 型柔性支链的静刚度模型表示为：

$$\boldsymbol{F}_{\mathrm{W}} = \boldsymbol{K}_{\mathrm{s}}\Delta\boldsymbol{P} \tag{1}$$

式中，$\boldsymbol{F}_{\mathrm{W}}$ 为终端外载荷的驱动力；$\boldsymbol{K}_{\mathrm{s}}$ 为支链的静刚度矩阵，包括机构传动刚度和柔性铰链刚度；$\Delta\boldsymbol{P}$ 为终端位移变形，使：$\boldsymbol{F}_{W} = (f_1 \quad f_2 \quad f_3)^{\mathrm{T}}, \Delta P = (p_1 \quad p_2 \quad p_3)^{\mathrm{T}}$。

将 RPU 柔性并联机构的支链仿真过程中的数值及结构代入上式中，可计算出 RPU 柔性并联机构支链的静刚度矩阵为：

$$K_{\mathrm{s1}} = \begin{bmatrix} -2.3850\mathrm{E}-01 & 0 & 0 \\ 0 & 1.4626\mathrm{E}-01 & 0 \\ 0 & 0 & 1.8428\mathrm{E}-02 \end{bmatrix}$$

3.1.6.3　RPU 全柔顺并联支链拓扑优化

应用 Ansys 软件[11]从 SolidWorks 中导入 RPU 型柔性并联机构支链模型，定义机构材料属性和单元类型，进行网格划分[12]。将支链上的一个支点设定自由度为 0，并在支链上施加驱动力[13]和外加载荷为 1000N。减少 80% 面积，消减后的伪密度的范围在 0.1～1 之间，同时进行 10 次累加迭代。变形后的图形如图 7 所示。

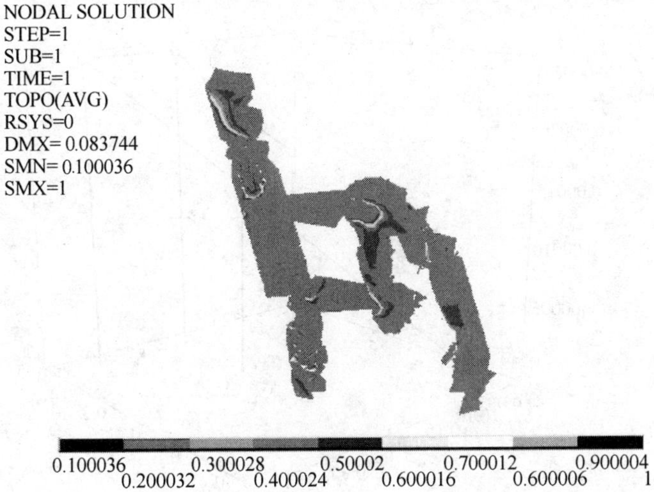

图 7　RPU 支链拓扑优化后的全柔顺并联支链形式

3.1.6.4 RPU 全柔顺并联支链刚度分析

在同等条件下对机构的 Ansys 模型进行计算和后处理，得到机构支链应变变形图如图 8 所示。

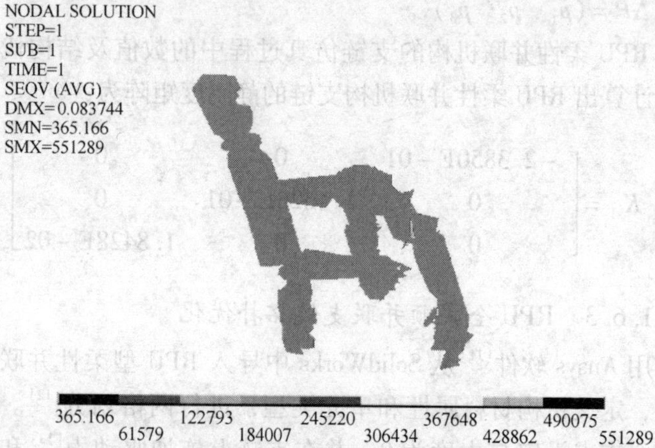

图 8 RPU 全柔顺并联支链应变变形图

选取支链输出平面上的若干关键节点，在外力作用下这些节点的位移变形如图 9 所示。

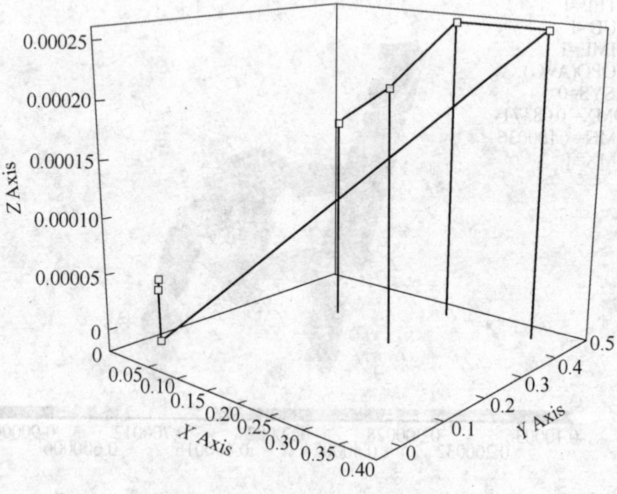

图 9 输出平面节点位移变形

对图 9 中的 UX、UY、UZ 分别求平均值，可得到位移：$\overline{UX}$ = 0.91172E－01，$\overline{UY}$ = 0.18202，$\overline{UZ}$ = 1.14794E－04，由此可知支链输出平面的运动。

RPU 型全柔顺并联支链的静刚度为：

$$\boldsymbol{K}_{s2}=\begin{bmatrix}91.172 & 0 & 0\\ 0 & 182.02 & 0\\ 0 & 0 & 1.14794\mathrm{E}-01\end{bmatrix}$$

3.1.7 SPS 型柔性及全柔顺并联机构刚度对比研究

3.1.7.1 SPS 柔性支链刚度分析

该支链应用 Ansys 软件分析刚度的设置与 RPU 型柔性支链的刚度分析设置完全相同。对机构的 Ansys 模型进行计算和后处理，得到机构支链总位移变形图如图 10 所示。

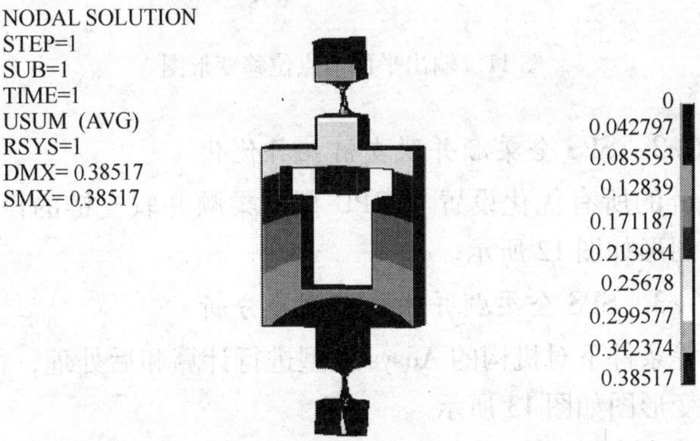

图 10 SPS 柔性支链的总位移变形图

选取支链输出平面上的若干关键节点，在外力作用下这些节点的位移变形如图 11 所示。

对上图中的 UX、UY、UZ 分别求平均值，可得到位移：$\overline{UX}$ = －0.38571E－02，$\overline{UY}$ = 0.24551，$\overline{UZ}$ = 0.22457E－01。

SPS 柔性支链的静刚度为：

$$\boldsymbol{K}_{s3} = \begin{bmatrix} -3.8570 & 0 & 0 \\ 0 & 245.51 & 0 \\ 0 & 0 & 22.457 \end{bmatrix}$$

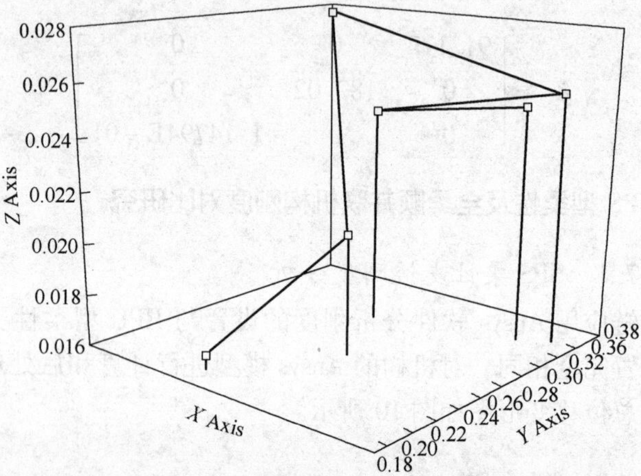

图 11 输出平面节点位移变形图

3.1.7.2 SPS 全柔顺并联支链拓扑优化

该支链的所有优化设置与 RPU 型全柔顺并联支链的设置相同，优化后的图形如图 12 所示。

3.1.7.3 SPS 全柔顺并联支链刚度分析

在同等条件下对机构的 Ansys 模型进行计算和后处理，得到机构支链应力变形图如图 13 所示。

选取支链输出平面上的若干关键节点，在外力作用下这些节点的位移变形如图 14 所示。

对上图中的 UX、UY、UZ 分别求平均值，可得到位移：$\overline{UX} = 0.22093$，$\overline{UY} = 0.23084\text{E}-01$，$\overline{UZ} = 0.23084\text{E}-01$。

SPS 型全柔顺并联支链的静刚度为：

$$K_{s4} = \begin{bmatrix} 220.93 & 0 & 0 \\ 0 & 337.81 & 0 \\ 0 & 0 & 23.084 \end{bmatrix}$$

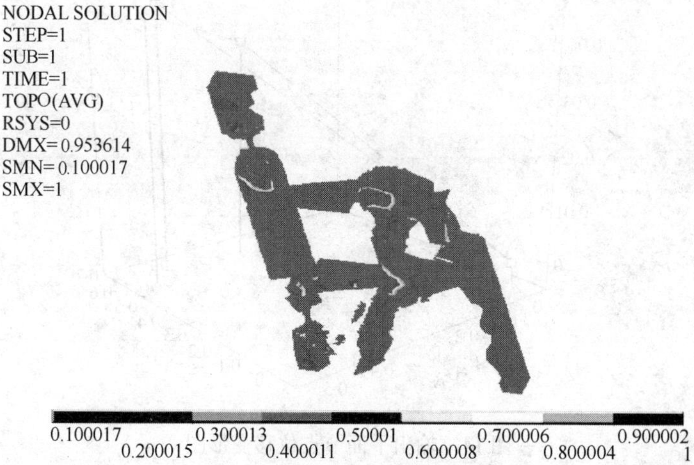

图 12　SPS 支链拓扑优化后的
全柔顺并联支链形式

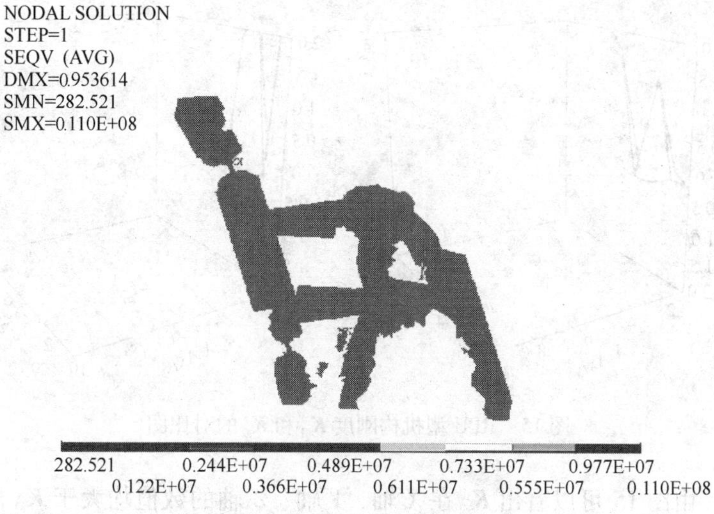

图 13　SPS 全柔顺并联支链应力变形图

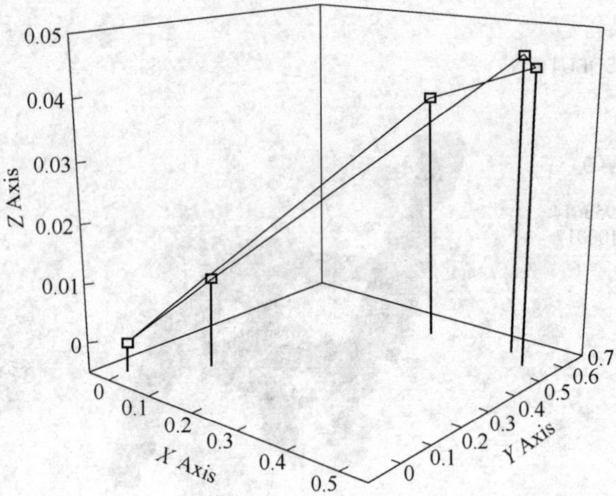

图 14 输出平面节点位移变形图

3.1.8 仿真对比研究

根据刚度 $\boldsymbol{K}_{s1}$ 和 $\boldsymbol{K}_{s2}$ 的数据，可得到对比图形如图 15 所示。

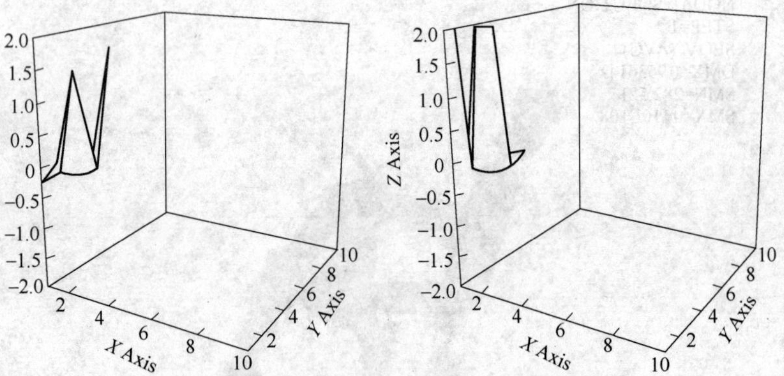

图 15 RPU 型机构刚度 $\boldsymbol{K}_{s1}$ 和 $\boldsymbol{K}_{s2}$ 的对比图

由图 15 可以看出 $\boldsymbol{K}_{s2}$ 在 X 轴、Y 轴、Z 轴的数值均大于 $\boldsymbol{K}_{s1}$，得出 RPU 型全柔顺并联支链的刚度大于 RPU 型柔性支链的刚度，从而证明全柔顺并联支链的设计方法在刚度方面更优。

SPS 柔性支链的刚度 $\boldsymbol{K}_{s3}$ 和 SPS 型全柔顺并联支链的刚度 $\boldsymbol{K}_{s4}$ 对比图如图 16 所示。

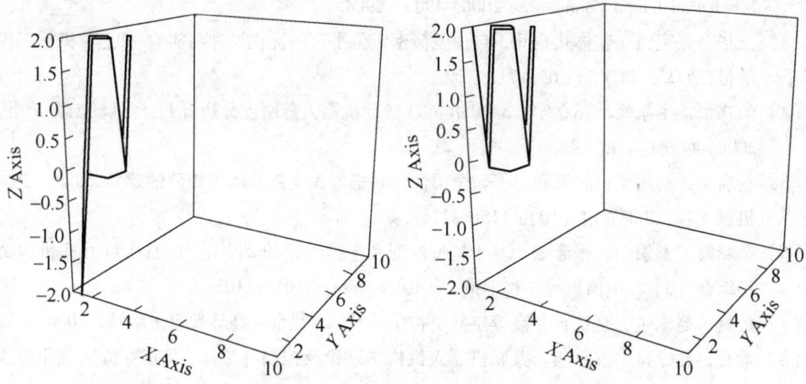

图 16 SPS 型机构刚度 $\boldsymbol{K}_{s3}$ 和 $\boldsymbol{K}_{s4}$ 的对比图

由图 16 可以看出 $\boldsymbol{K}_{s4}$ 的最小值远远大于 $\boldsymbol{K}_{s3}$ 的最小值；$\boldsymbol{K}_{s4}$ 的变化范围远远小于 $\boldsymbol{K}_{s3}$ 的范围，从而得出 RPU 型全柔顺并联支链的刚度大于 RPU 型柔性支链的刚度，从而证明全柔顺并联支链的设计方法在刚度方面更优。

3.1.9 结论

本文对 2RPU-2SPS 型全柔顺并联机构进行分析，根据并联机构设计出柔性并联机构，然后在此基础上设计出全柔顺并联机构。采用 Ansys 软件对全柔顺并联机构的支链进行拓扑优化设计，综合各方面因素，得出全柔顺并联机构各支链构型。分析柔性并联机构和全柔顺并联机构支链的刚度，并对结果进行对比研究，得出：（1）全柔顺并联机构的支链刚度大于柔性支链的刚度；（2）全柔顺并联机构设计的必要性。本研究为全柔顺并联机构的整体设计提供了理论依据。

参 考 文 献

[1] Lee Jae-Jong, Choi Kee-Bong, KimGee-Hong. Design and analysis of the single-step nanoim printing lithography equipment for sub-100 nm linewidth [J]. Current Applied Physics, 2006

(6): 1007 ~ 1011.

[2] Lee Jae-Jong, Choi Kee-Bong, KimGee-Hong, et al. The UV-nanoim print lithography with multi-head nanoim printing unit for sub-50nm half pitch patterns [C] //Korea: SICEICASE International Joint Conference, 2006: 4902 ~ 4904.

[3] 王华，张宪民. 整体式空间3 自由度精密定位平台的优化设计与实验 [J]. 机械工程学报, 2007, 43 (3): 65 ~ 71.

[4] 荣伟彬，王乐峰，孙立宁. 3-PPSR 并联微动机器人静刚度分析 [J]. 机械工程学报, 2008, 44 (1): 13 ~ 24.

[5] 杨启志，马履中，郭宗和，等. 全柔性并联机器人支链静刚度矩阵的建立 [J]. 中国机械工程, 2008, 19 (10): 1156 ~ 1159.

[6] 范彩霞，刘宏昭，张彦斌. 基于构型演变和李群理论的 2T2R 型四自由度并联机构构型综合 [J]. 中国机械工程学报, 2010, 5 (9): 1101 ~ 1105.

[7] 黄真，赵永生，赵铁石. 高等空间机构学 [M]. 北京：高等教育出版社, 2006.

[8] 黄真，孔令富，方跃法. 并联机器人机构学理论及控制 [M]. 北京：机械工业出版社, 1997.

[9] 刘宏伟. 基于螺旋理论的少自由度并联机构运动学分析 [J]. 制造业自动化, 2009, 31 (7): 101 ~ 103.

[10] 江涛，朱大昌. 空间三平移柔顺并联超精密定位平台设计与分析 [J]. 煤矿机械, 2011, 32 (10): 49 ~ 52.

[11] 高耀东，刘学杰. Ansys 机械工程应用精华 50 例 [M]. 北京：电子工业出版社, 2011.

[12] 周文闻，何广平. 平面全柔性并联机构柔性铰链的优化分析 [J]. 北方工业大学学报, 2007, 19 (1): 20 ~ 27.

[13] 梁毓明，陈德海. 轮式移动机器人调速系统的设计 [J]. 江西理工大学学报, 2008, 29 (4): 13 ~ 16.

3.2 四自由度全柔顺并联机构刚度分析

本文利用动力学方程求出 4-CRU 型全柔顺并联机构支链的刚度，并用 Ansys 软件仿真结果验证此方法的正确性，为求全柔顺并联机构整体刚度提供了依据；本文基于 4-CRU 并联机构的运动特性，采用传统柔性并联机构的构成方式设计了具有全柔顺特性的新型 4-CRU 型全柔顺并联机构；利用动力学方程对全柔顺并联支链刚度进行计算；采用 Ansys 软件建立 4-RCU 型全柔性并联机构支链模型，并对其刚度进行仿真实验对比研究；仿真结果与计算结果基本上一致，验证了分析方法的正确性。

3.2.1 引言

相对于串联机构，并联机构具有无误差累计、运动惯量小、刚度大、反解较简单等优点，更适合用于开发精密定位平台，如六自由度的 Stewart 飞行模拟器平台[1]。由于并联机构各运动副之间装配误差及存在的间隙对末端运动平台运动特性所造成的影响，以柔性铰链代替传统运动副的柔性并联机构被提出。该机构利用柔性元件的弹性变形而非刚性元件的相对运动来转换或传递运动、力和能量，是一种具有并联机构运动特性的新型免装配机构[2]。其具有免装配、无间隙、无摩擦及免润滑等诸多优点，将并联机构支链上各运动副及连杆以柔性机构代替。近年来，针对柔性并联机构的研究很多，但大部分研究都只是将并联机构各运动支链的运动副以柔性铰链代替，再用柔性连杆将各个柔性铰链连接在一起形成相对应的柔性支链，从而与运动平台及固定基座一起构成柔性并联机构，即替换法。如杨启志、马履中等[3]设计的一种非对称全柔性 3-RRRP 平移微动并联机构，其在非对称 3-RRRP 型并联机构的基础上，将并联机构的所有运动副柔性化，形成柔性铰链，从而构成全柔性 3-RRRP 并联机构。通过这种方法设计的柔性并联机构可实现所要求的运动与功能，但在一些具有更高精度要求的领域上，柔性并联机构要求具有更高的刚度且具有更佳的运动特性。本文将柔性机构和并联机构有机结合，设计出具有四自由度的 4-CRU 型全柔顺并联机构。这种柔顺并联机构与柔性并联机构相比，其最大的不同在于其各支链均是通过先进切割技术整体切割而成，这些支链可称为全柔顺支链。由于其是整体式构型，所以理论上它比柔性并联机构具有更高的刚度及运动精度。因此，对全柔顺并联机构进行设计及研究，并以此为基础设计相应的精密定位平台，具有重要的理论研究意义和巨大的实用价值。由于刚度分析是对这种新的机构的分析，本身就与以往的不同。此类新型机构在工作空间、奇异性等方面还需要做大量的研究。

3.2.2 四自由度并联机构的运动特性

文中研究的具有四自由度的 4 -CRU 型并联机构的结构图如图 1

所示。文中对其运动特性的具体分析如下。

在分析并联机器人机构中，螺旋理论是一个非常有效的工具[4,5]。螺旋的基本形式可以表示为：$\hat{\$}=[\boldsymbol{s} \quad \boldsymbol{s}_0]^{\mathrm{T}}$，其中，$\boldsymbol{s}$ 为沿螺旋轴线方向的单位矢量，s_0 为对偶数，当螺旋 $\$$ 的两矢量表示为标量时，可用 *plcüker* 坐标(l,m,n,o,p,q)来表示。如果有两个螺旋 $\$$ 和 $\$_r$，若满足下述条件：$\$ \circ \$_r=0$，则 $\$$ 和 $\$_r$互为反螺旋，式中"∘"表示互易积。从物理意义上讲，互易积为零的两个螺旋，一个可用来表示物体的运动，另一个则表示物体受力情况。零节距的力螺旋描述为力，无穷大节距的力螺旋表示为力偶。

如图 2 所示，在动平台上建立基坐标系 $B-XYZ$，X 轴和 Y 轴在动平台内，Z 轴垂直于动平台。在定平台上建立支链 I 的坐标系 $B_1-x_1y_1z_1$，其中，z_1 垂直于定平台，x_1 轴与动平台上移动副 B_1 的轴线平行。

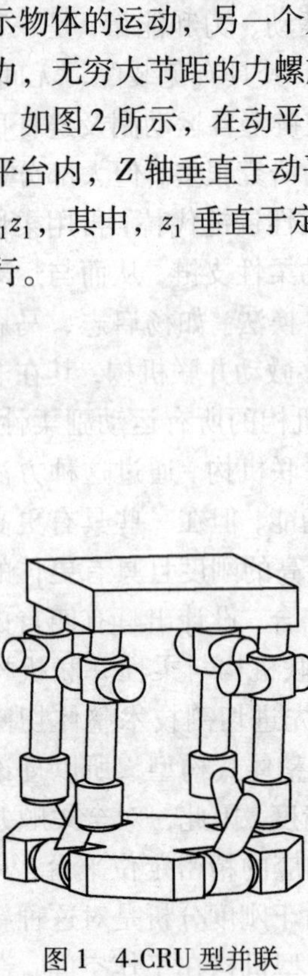

图 1　4-CRU 型并联机构结构图

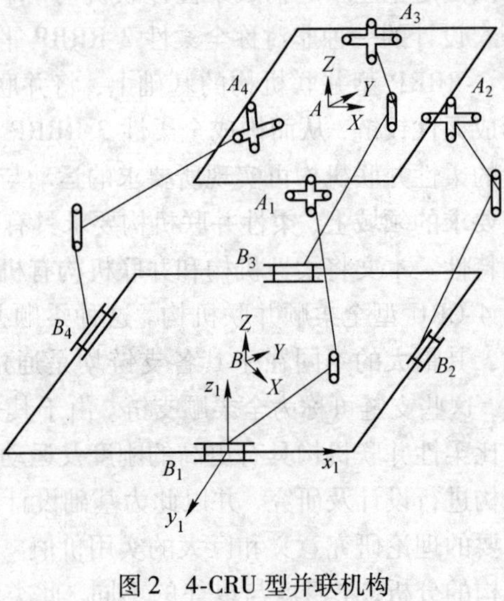

图 2　4-CRU 型并联机构坐标系建立

支链 I 中 $\$_{11}$ 与 $\$_{12}$同轴，$\$_{11}//o_1x_1$，$\$_{13}//o_1z_1$，$\$_{14}//o_1x_1$，$\$_{15}//o_1z_1$，支链 I 的运动螺旋为：

$$\$_{11}=(1,0,0,0,0,0),\ \$_{12}=(0,0,0,1,0,0),\ \$_{13}=(0,0,1,0,0,0),$$
$$\$_{14}=(1,0,0,0,0,0),\ \$_{15}=(0,0,1,0,0,0) \tag{1}$$

其约束反螺旋为：

$$\$_1^r=(0,0,0,1,1,0) \tag{2}$$

从支链 I 的反螺旋系可以看出，支链 I 对动平台提供一个平行于 *XOY* 平面的约束力偶，4 个分支都具有相同的结构。4 个分支提供 4 个约束力偶，且都平行于 *XOY* 平面汇交，故机构具有绕 *Z* 轴的转动和沿 *X*、*Y*、*Z* 轴的移动。

3.2.3 四自由度全柔顺并联机构构型设计

在具有四自由度的 4-CRU 型并联机构的基础上设计与之相对应的 4-CRU 型四自由度全柔顺并联机构[6]。首先，设计相应的 CRU 型全柔顺支链；其次，全柔顺支链均以相邻支链垂直且均匀地固定在定平台上的方式装配起来，即形成 4-CRU 型四自由度全柔顺平台。全柔顺支链（如图 3 所示）是在选定一整块合适的材料上通过先进切割技术分别切割出各个相应的柔性铰链和柔性连杆而制成。CRU 型全柔顺支链的切割是按要求切割出两个柔性转动副 R 4、一个四连杆型柔性移动副 P（由两个转动副 R5 组成）、一个柔性虎克铰 U（由轴线互相垂直的 R 1 和 R 2 组成）、一个柔性转动副 R 3 和连接各个柔性铰链的柔性连杆。在常规铰链中，圆柱副可以由一个转动副和一个移动副构成，因此，在 CRU 型全柔顺支链中设计两个柔性转动副 R 5 和四连杆型柔性移动副 P 构成了一个柔性圆柱副 C。

用 4 个 CRU 型全柔顺支链将固定基座和运动平台连接在一起就构成了具有四自由度的 4 -CRU 型全柔顺并联机构。其中，运动平台固定在 4 个 CRU 型全柔顺支链的运动输出端上，如图 4 所示。

3.2.4 四自由度全柔顺并联机构支链刚度研究

3.2.4.1 系统假设条件

为了便于计算做出如下假设[7]：

（1）连续性假设：认为组成固体的物质毫无间隙地充满固体的几何空间；

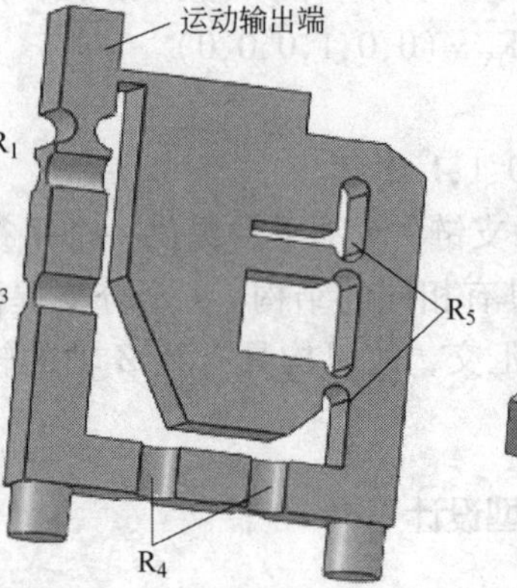

图 3 4-CRU 型全柔顺支链

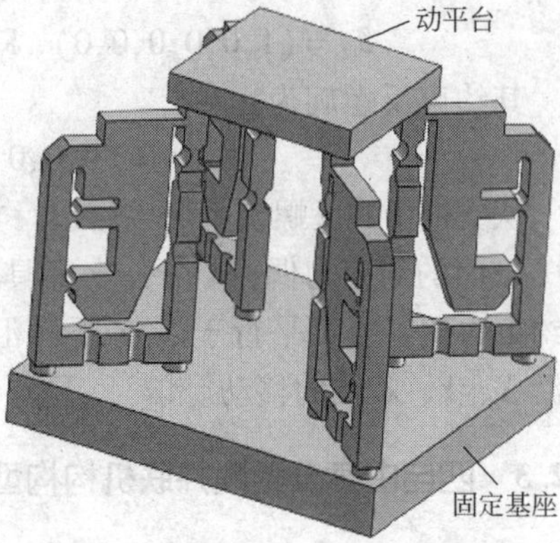

图 4 4-CRU 型全柔顺并联机构结构图

（2）弹性小变形条件假设：固体因外力作用而引起的变形（限于变形的大小）远远小于铰链的原始尺寸的情况，且在外力消除后又可恢复原状；

（3）不考虑各构件变形：动平台比各支链杆的尺寸大得多，其变形远远小于柔性铰链的变形，因此，假设动平台为刚体。

3.2.4.2 系统单元划分和空间单元模型的建立

动平台 $A_1A_2A_3A_4$ 设为单元 0，每个铰链看做一个单元，即 $i_1 \sim i_2$（$i=1, 2, 3, 4$），这样每个支链设为 7 个柔性单元，系统共计有 28 个单元。选择矩形截面空间梁单元作为基本梁单元模型，如图 5 所示。此空间梁单元[8]划分了 2 个节点，分别标记为 A、B。$\delta_1 \sim \delta_3$ 与 $\delta_{10} \sim \delta_{12}$，$\delta_4 \sim \delta_6$ 与 $\delta_{13} \sim \delta_{15}$，$\delta_7 \sim \delta_9$ 与 $\delta_{16} \sim \delta_{18}$ 分别表示节点 A、B 处的弹性位移、弹性转角和曲率。固定在单元上的坐标系 $o-xyz$ 称为单元坐标系，规定取 $A-B$ 方向为 y 的正方向，把 y 按逆时针方向转过 90°的方向为 z 的正方向，x 同时垂直于 y 和 z。采用单元坐标分析单元节点变形和节点力间的关系更为简便。

3.2.4.3 单元位移函数

柔性铰链采用梁单元进行离散，节点数可以按精度要求任意选择。为描述简便，图 5 中只画出 2 个节点（分别标记为 A 和 B），文中推导过程参照 2 个节点的情形表述。假设空间梁单元发生轴向、横向（两个方向）、扭转变形，并用 $\boldsymbol{\delta}=[\delta_1,\ \delta_2,\ \cdots,\ \delta_{18}]^{\mathrm{T}}$ 表示梁单元的广义坐标向量，其中各分量表示单元端点的弹性位移、弹性转角和曲率。这样单元上任意点的相对于单元坐标系产生的沿 x、y、z 轴的弹性位移和绕 x、y、z 轴的弹性角位移等，皆可表示为 δ 的函数。

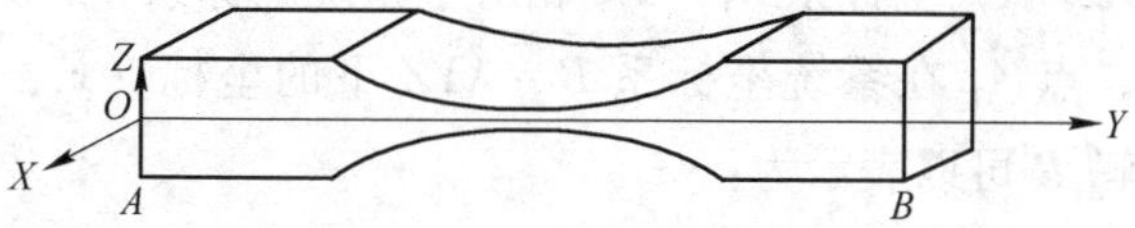

图 5　空间柔性铰链单元模型

3.2.4.4　单元动力学方程

根据 Lagrange 方程，得出单元铰链的动力学方程：

$$\boldsymbol{M}_{\mathrm{e}}\ddot{\boldsymbol{\delta}}+\boldsymbol{K}_{\mathrm{e}}\boldsymbol{\delta}=\boldsymbol{F}_{\mathrm{e}}+\boldsymbol{P}_{\mathrm{e}}+\boldsymbol{Q}_{\mathrm{e}} \tag{3}$$

式中，$\boldsymbol{M}_{\mathrm{e}}$ 为单元质量矩阵；$\boldsymbol{K}_{\mathrm{e}}$ 为单元刚度矩阵；$\boldsymbol{F}_{\mathrm{e}}$ 为单元外加载荷的广义列阵；$\boldsymbol{P}_{\mathrm{e}}$ 与研究的梁单元相连接的其他单元给予研究单元的作用力列阵；$\boldsymbol{Q}_{\mathrm{e}}$ 为系统单元刚度惯性力列阵。

根据文献［9］可知单元铰链的刚度矩阵为：

$$\boldsymbol{K}_{\mathrm{e}}=\begin{bmatrix} A & 0 & 0 & -A & 0 & 0\\ 0 & B & C & 0 & -B & C\\ 0 & C & D & 0 & -C & E\\ -A & 0 & 0 & A & 0 & 0\\ 0 & -B & -C & 0 & B & C\\ 0 & C & E & 0 & -C & E \end{bmatrix}$$

式中，

$$A=\frac{6Ebtt_4 4t_1}{tt_1+t_4t_1+4tt_4},\ B=\frac{8Eb_4^3t_1^3}{3l^3(2t_1^3+t_4^3)},\ C=\frac{4Eb_4^3t_1^3}{3l^3(2t_1^3+t_4^3)},$$

$$D=\frac{Eb(2t_4^3+4t_1^3t_4^3+5t_1^3)t_4^3t_1^3}{3l(t_4^3t_1^3+t^3t_1^3+2t^3t_4^3)(2t_1^4+t_4^3)},\ E=\frac{Eb(2t_4^3t_1^3+3t^3t_4^3)t_4^3t_1^3}{3l(t_4^3t_1^3+t^3t_1^3+2t^3t_4^3)(2t_1^4+t_4^3)}$$

3.2.4.5 支链的运动学约束

由于操作任务的需要，一般变形主要来源于柔性铰链，其余连接处的变形可以忽略不计。空间的一个刚体具有 6 个独立的自由度，所以支链与各个铰链的节点不是独立的，它们是支链 6 个独立参量的函数，在连接两个铰链时必须满足同一刚体上位移一致的条件。根据这个条件，进行支链运动学约束关系的推导。

如图 6 所示，坐标系 A_1-xyz 相对于系统坐标系 $B-XYZ$ 的变换矩阵为 ${}_{A_1}^{B}\boldsymbol{R}$，点 A_1 在系统坐标系 $B-XYZ$ 下的坐标 $(x_A, y_A, z_A)^{\mathrm{T}}$，则变换矩阵 ${}_{A_1}^{B}\boldsymbol{R}$ 可以表示为：

$$
{}_{A_1}^{B}\boldsymbol{R}=\begin{bmatrix} c\alpha c\beta & c\alpha s\beta s\gamma - s\alpha c\gamma & c\alpha s\beta c\gamma + s\alpha c\gamma & x_A \\ s\alpha c\beta & s\alpha s\beta s\gamma + c\alpha c\gamma & s\alpha s\beta c\gamma - c\alpha s\gamma & y_A \\ -s\beta & c\beta s\gamma & c\beta c\gamma & z_A \\ 0 & 0 & 0 & 1 \end{bmatrix}
$$

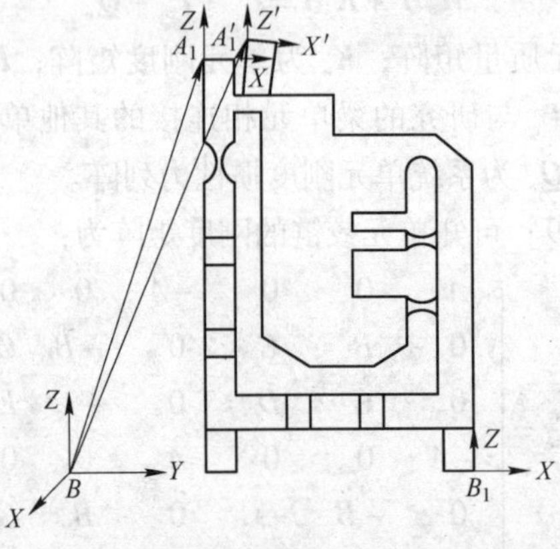

图 6 A_1 与支链的约束关系

假设支链的名义运动姿态位置在点 A_1 处，由于支链的各柔性铰链的弹性变形使得 A_1 的实际运动姿态位置发生微小变化（即 $\delta\alpha$，$\delta\beta$，$\delta\gamma$，δx_A，δy_A，δz_A）最终移动到点 A_1' 处。由坐标系 $A_1'-x'y'z'$ 到坐标系 A_1-xyz 的变换矩阵为 $\Delta\boldsymbol{R}$，其近似表达式为：

$$\Delta \boldsymbol{R}=\begin{bmatrix}1 & -\delta\alpha & -\delta\beta & \delta x_A\\ \delta\alpha & 1 & -\delta\gamma & \delta y_A\\ -\delta\beta & \delta\gamma & 1 & \delta z_A\\ 0 & 0 & 0 & 1\end{bmatrix} \tag{4}$$

由坐标系 $A_1'-x'y'z'$ 到系统坐标系 $B-XYZ$ 的变换矩阵为 ${}_{A_1}^{B}\boldsymbol{R}$，系统坐标系 $B-XYZ$ 到 B_1-xyz 的变换矩阵为 $\boldsymbol{T}$，则坐标系 $A_1'-x'y'z'$ 到坐标系 B_1-xyz 的变换矩阵 $\boldsymbol{T}_1$ 为：

$$\boldsymbol{T}_1={}_{A_1}^{B}\boldsymbol{R}\boldsymbol{T}=\Delta\boldsymbol{R}_{A_1}^{B}\boldsymbol{R}\boldsymbol{R}(\varphi) \tag{5}$$

式中，φ 为不同值，当 $i=1$，$\varphi=0$，$i=2$，$\varphi=90^\circ$，$i=3$，$\varphi=180^\circ$，$i=4$，$\varphi=270^\circ$。

设图 6 中的点 A_1 和 A_1' 在坐标系 B_1-xyz 下的坐标分别为（x_{A_1}，y_{A_1}，z_{A_1}）^{T} 和（x'_{A_1}，y'_{A_1}，z'_{A_1}）^{T}，则有：

$$\begin{bmatrix}\Delta x_{A_1}\\ \Delta y_{A_1}\\ \Delta z_{A_1}\end{bmatrix}=\begin{bmatrix}1 & 0 & 0 & 0 & z_{A_1} & -y_{A_1}\\ 0 & 1 & 0 & -z_{A_1} & 0 & x_{A_1}\\ 0 & 0 & 1 & y_{A_1} & -x_{A_1} & 0\end{bmatrix}\begin{bmatrix}\delta x_{A_1}\\ \delta y_{A_1}\\ \delta z_{A_1}\\ \delta\gamma\\ \delta\beta\\ \delta\alpha\end{bmatrix} \tag{6}$$

由式（7）可以得到由 $\boldsymbol{U}_{A_1}$ 和 $\boldsymbol{U}_{B_1}$ 表示 A 与支链之间的运动学约束为：

$$\boldsymbol{U}_{A_1}=\begin{bmatrix}1 & 0 & 0 & 0 & z_{A_1} & -y_{A_1}\\ 0 & 1 & 0 & -z_{A_1} & 0 & x_{A_1}\\ 0 & 0 & 1 & y_{A_1} & -x_{A_1} & 0\end{bmatrix}\boldsymbol{U}_{B_1} \tag{7}$$

或简记为：

$$\boldsymbol{U}_{A_1}=\boldsymbol{R}_1\boldsymbol{U}_{B_1} \tag{8}$$

3.2.4.6 支链的刚度分析

单元坐标下的支链动力学方程为：

$$\boldsymbol{M}_1\ddot{\boldsymbol{\delta}}_0+\boldsymbol{K}_1\boldsymbol{\delta}_0=F_1+P_1+Q_1 \tag{9}$$

因 A_1 的位移改变量是由各铰链的改变量引起的，所以 A_1 在单元坐标下的位移改变量与系统坐标下的改变量关系为：

$$\boldsymbol{\delta}_0 = \boldsymbol{B}_0 \boldsymbol{U}_{A_1} \tag{10}$$

式中，$\boldsymbol{B}_0 = \begin{bmatrix} \boldsymbol{R}_0 & 0 \\ 0 & \boldsymbol{R}_0 \end{bmatrix}$；$\boldsymbol{\delta}_0 = [\delta_{01} \quad \delta_{02} \quad \delta_{03} \quad \delta_{04} \quad \delta_{05} \quad \delta_{06}]$ 为各个铰链的变形引起的 A_1 的位移改变量在 A_1-xyz 下的表示；$\boldsymbol{U}_{A_1} = [U_{01} \quad U_{02} \quad U_{03} \quad U_{04} \quad U_{05} \quad U_{06}]$ 为各个铰链的变形引起的 A_1 的位移改变量在 $B-XYZ$ 下的表示。由各个铰链变形引起的 A_1 位移改变量 $\boldsymbol{R}_0 = \begin{bmatrix} 1 & 0 & 0 \\ 0 & 1 & 0 \\ 0 & 0 & 1 \end{bmatrix}$ 为系统坐标表示到单元坐标的姿态变换矩阵。

将式（10）代入式（9）中，用矩阵左乘可得：

$$R_1^{\mathrm{T}} M_1 \boldsymbol{B}_0 \boldsymbol{R}_1 \ddot{\boldsymbol{U}}_{B_1} + \boldsymbol{R}_1^{\mathrm{T}} \boldsymbol{K}_e \boldsymbol{B}_0 \boldsymbol{R}_1 \boldsymbol{U}_{B_1} = R_1^{\mathrm{T}}(F_1 + P_1 + Q_1) \tag{11}$$

令 $\boldsymbol{M} = R_1^{\mathrm{T}} M_1 \boldsymbol{B}_0 \boldsymbol{R}_1$，$\boldsymbol{K} = \boldsymbol{R}_1^{\mathrm{T}} \boldsymbol{K}_e \boldsymbol{B}_0 \boldsymbol{R}_1$，$\boldsymbol{F} = R_1^{\mathrm{T}}\ (F_1 + P_1 + Q_1)$，则式（11）可变形为：

$$\boldsymbol{M}\ddot{\boldsymbol{U}}_{B_1} + \boldsymbol{K}\boldsymbol{U}_{B_1} = \boldsymbol{F} \tag{12}$$

式中，$\boldsymbol{M}$ 为支链质量矩阵；$\boldsymbol{K}$ 为支链的刚度矩阵；$\boldsymbol{F}$ 为力列阵。

由 Ansys 分析处支链上 A_1 点的运动位移，已知条件为：弹性模量 $E = 3.5 \times 10^5 \mathrm{N/m^2}$，泊松比 $\mu = 0.499$，剪切弹性模量 $G = 1.167 \times 10^5 \mathrm{N/m^2}$，单个铰链参数为：$l = 0.015\mathrm{m}$，$t = 0.005$，$h = 0.002$，得出支链的刚度矩阵为：

$$\boldsymbol{K} = \begin{bmatrix} 6.639 \times 10^{-2} & 0 & -0.0016 \\ 0.0002 & -1.31 \times 10^{-3} & 0 \\ -0.0016 & 0 & 6.667 \times 10^{-2} \end{bmatrix}$$

3.2.5 四自由度全柔顺并联机构支链刚度 Ansys 分析

3.2.5.1 四自由度全柔顺并联机构仿真分析

应用 Ansys 软件[10] 从 SolidWorks 中导入 4-CRU 型四自由度全柔顺并联机构的支链模型，定义机构材料属性和单元类型，进行网格划分。将支链上的一个支点设定自由度为0，并在支链上施加驱动力和外加载荷为200N。支链的网格划分和总的变形图如图7及图8所示。

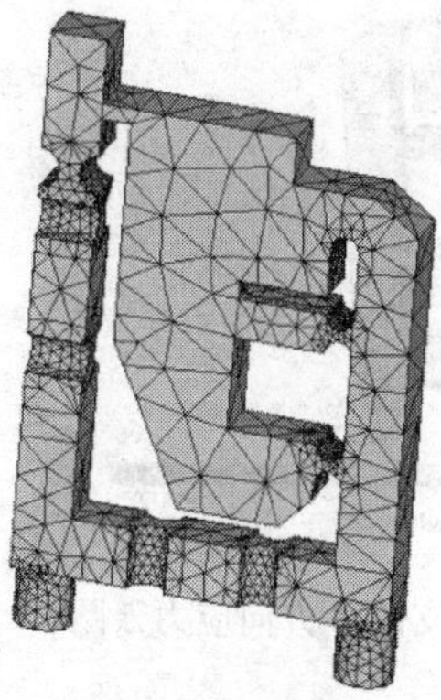

图7 四自由度全柔顺并联机构支链网格划分

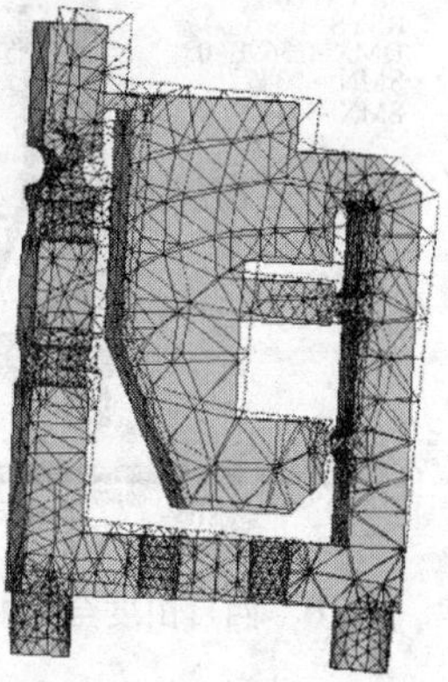

图8 四自由度全柔顺并联机构支链总变形图

对机构的Ansys模型进行计算和后处理，得到机构支链总位移变形图和应力分布云图，如图9和图10所示。

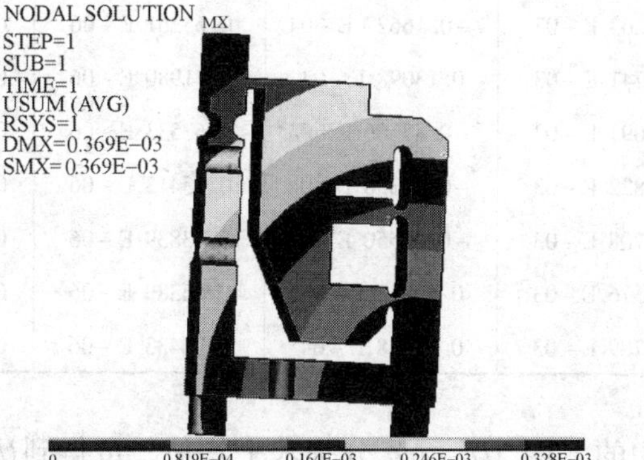

图9 四自由度全柔顺并联机构支链总位移图

选取支链输出平面上的若干关键节点，在外力作用下这些节点的位移变形如表1所示。

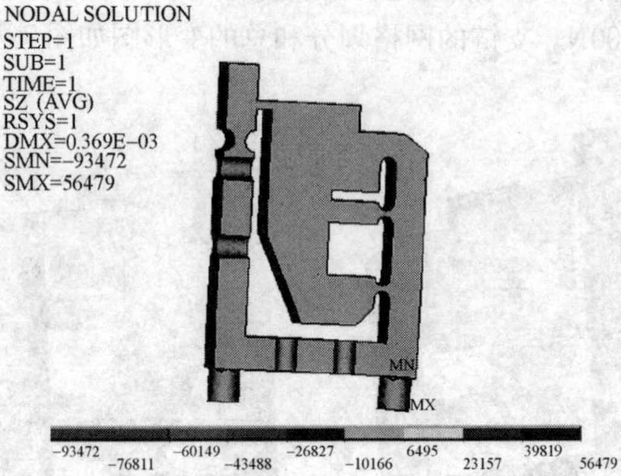

图 10 四自由度全柔顺并联支链 Z 方向应力云图

表 1 输出平面节点位移变形

Node	*UX*	*UY*	*UZ*	*USUM*
33	0. 36779E − 03	0. 24149 E − 04	0. 74575 E − 06	0. 36859 E − 03
96	0. 35128 E − 03	− 0. 15230 E − 04	0. 70549 E − 06	0. 35161 E − 03
438	0. 31551 E − 03	− 0. 15897 E − 04	0. 73552 E − 06	0. 31591 E − 03
942	0. 33263 E − 03	− 0. 16627 E − 04	0. 65261 E − 06	0. 33305 E − 03
2609	0. 35741 E − 03	0. 13098 E − 04	0. 71980 E − 06	0. 35765 E − 03
2805	0. 33091 E − 03	− 0. 83106 E − 05	0. 67533 E − 06	0. 33101 E − 03
4098	0. 29822 E − 03	− 0. 34338 E − 04	0. 65413 E − 06	0. 30019 E − 03
4152	0. 30708 E − 03	− 0. 30550 E − 04	0. 73839 E − 06	0. 30860 E − 03
10759	0. 30576 E − 03	0. 47832 E − 05	0. 66389 E − 06	0. 30580 E − 03
14354	0. 35789 E − 03	0. 13438 E − 04	0. 76433 E − 06	0. 35814 E − 03

对表 1 中的 UX、UY、UZ 分别求平均值，可得到 $\overline{UX} = 3.32\text{E}-04$，$\overline{UY} = -6.5\text{E}-06$，$\overline{UZ} = 3.33\text{E}-04$

3.2.5.2 四自由度全柔顺并联机构刚度计算

四自由度全柔顺并联机构的静刚度模型表示为：

$$\boldsymbol{F}_{\mathrm{w}} = \boldsymbol{K}_{\mathrm{s}} \Delta \boldsymbol{P} \tag{13}$$

式中，$\boldsymbol{F}_{\mathrm{w}}$ 为终端外载荷驱动力；$\boldsymbol{K}_{\mathrm{s}}$ 为支链静刚度矩阵，包括机构传动刚度和柔性铰链刚度；$\Delta \boldsymbol{P}$ 为终端位移变形，有：

$$\boldsymbol{F}_{\mathrm{w}} = (f_1 \quad f_2 \quad f_3)^{\mathrm{T}},\ \Delta P = (p_1 \quad p_2 \quad p_3)^{\mathrm{T}} \tag{14}$$

将4-CRU型四自由度全柔顺并联机构支链仿真过程中的数值及结果代入式（14）中，由式（13）可得到4-CRU型四自由度全柔顺并联机构支链的静刚度矩阵为：

$$\boldsymbol{K}_{\mathrm{s}} = \begin{bmatrix} 6.64 \times 10^{-2} & 0 \\ 0 & -1.3 \times 10^{-3} \\ 0 & 0 \end{bmatrix}$$

3.2.6 结论

本文通过采用两种方法对4-CRU型四自由度全柔顺并联机构刚度矩阵进行计算与分析，结果表明通过两种方法对刚度矩阵先通过求出单个铰链的刚度矩阵，然后再整合整个支链刚度矩阵的方法是可行的，该方法为用动力学方程求出整个全柔顺并联机构的刚度矩阵提供了理论计算依据，为超精密定位平台的振动抑制研究提供了理论研究依据。

参 考 文 献

[1] Seling J M, Ding X. Theory of vibration in Stewart platforms [C] //Proceedings of the 2001 International Conference on Intelligent Robot and Systems, Maui, Hawaii, USA, 2001: 2190~2195.

[2] Dado M H F. Limit position synthesis and analysis of compliant 4-bar mechanisms with specified energy levels using variable parmetric pseudo-rigid-body models [J]. Mechanism and Machine Theory, 2005, 40 (8): 977~992.

[3] 杨启志，马履中，郭宗和等. 全柔性并联机器人支链静刚度矩阵的建立 [J]. 中国机械工程, 2008, 19 (10): 1156~1159.

[4] 黄真，赵永生，赵铁石. 高等空间机构学 [M]. 北京：高等教育出版社, 2006.

[5] 黄真，孔令富，方跃法. 并联机器人机构学理论及控制 [M]. 北京：机械工业出版社, 1997.

[6] 王华，张宪民，邓俊广. 基于压电陶瓷驱动的精密定位平台研究 [J]. 测试技术学报, 2007, 21 (4): 295~300.

[7] 周玉林，高峰. 2-RRR + RRS 球面并联机构的静刚度分析 [J]. 机械设计与研究,

2008, 24 (4): 35 ~ 40.
[8] 刘善增. 三自由度空间柔性并联机器人动力学研究 [D]. 北京: 北京工业大学, 2009.
[9] 尹小琴, 马履中, 杨启志等. 3-RRC 全柔性机构中柔性铰链刚度矩阵建立 [J]. 江苏大学学报 (自然科学版), 2003, 24 (4): 6 ~ 8.
[10] 高耀东, 刘学杰. Ansys 机械工程应用精华 50 例 [M]. 北京: 电子工业出版社, 2011.

3.3 Structural Design of a 3-DoF UPC Type Rotational Fully Spatial Compliant Parallel Manipulator

Conventionally, spatial compliant parallel manipulators are designed by replacing a rigid kinematic manipulator pair with flexible joints. Replacement of the manipulator pair is accomplished at a cost of the modal stiffness, which is the main influencing factor in vibration suppression. In this paper, the geometric constraint conditions used for the enumeration of the feasible limb structures for a class of 3-degree-of-freedom (DoF) rotational, fully compliant parallel manipulators is derived using the theory of reciprocal screws. As an example, a 3-DoF UPC-type rotational parallel manipulator is used to reconfigure 3-DoF rotational, fully spatial compliant parallel manipulators. This method is based on the topology optimization theory for a set of geometric constraint conditions. Compared with a conventional compliant parallel manipulator, the simulation results show the validity of the method for designing a fully spatial compliant parallel manipulator with high modal stiffness and vibration suppression.

3.3.1 Introduction

Parallel manipulators have attracted a significant amount of attention among researchers and engineers in the past two decades because they possess inherent advantages over conventional serial manipulators, such as high rigidity, load capacity, velocity, and precision. However, as for any mechanical system composed of conventional joints, traditional parallel manipulators suffer errors due to backlash, hysteresis, and manufacturing flaws in the joints. Hence, it is a major challenge to achieve ultra-high precision using conventional joints. Piezoelectric actuating techniques become some of the most important techniques in the fields of micromanipulation and microposition, compliant mechanisms have attracted more attention. Compliant mechanisms are flexible structures that generate some desired motions through elastic deformation instead of through rigid linkages, as in rigid body mechanisms. Compliant mechanisms are suitable for use in the field of precision engineering, as a result of the monolithic nature of the compliant mechanisms with few or no rigid body joints. The lack of rigid body joints serves to reduce friction and backlash losses, part assembly costs, noise and vibrations and facilitates unitization[1,2].

In particular, 3-DoF parallel manipulators with translational or rotational moving platforms have attracted attention because of their potential in applications such as positioning or orienting devices[3~11]. The geometrical constraint conditions are such that each limb in a parallel mechanism imposes one or more constraints on the moving platform, and the mobility of the moving platform is determined by the intersection of these constraints. It is therefore natural to consider screw theory for the structural synthesis of such mechanisms[12,13]. A few parallel manipulators employing compliant mechanisms have been designed to perform manipulation on a micro/nanometer scale with high accuracy, speed and load capacity. Minoru[14] constructed a compliant wrist with a parallel link based on passive compliance, Cartesian stiffness and suitable work space. Byoung[15] considered the analysis and design of a general platform type of parallel mechanism containing flexure joints. From this work, the Pareto frontier is obtained, which can be used to select the desired design parameters based on secondary criteria, such as performance sensitivity. Yangmin Li[16] proposed a 3-DoF translational compliant parallel platform and established a pseudo-rigid-body model. Qingsong Xu[17] proposed a compliant parallel micromanipulator with a 3-PRC architecture, as shown in Fig. 1.

Huang[18] presented a completely decoupled *XY* micro-motion stage, which is a compliant parallel mechanism based on flexure hinges and driven by piezoelectric actuators. In this study, the workspace is analyzed by considering the output reduction of PZT, stress in the flexure hinges and buckling. Additionally, a novel two-DoF compliant parallel micromanipulator driven by PZT was presented based on flexure hinges[19]. Boudewijn[20] presented the design, modeling and fabrication of a planar three-DoF parallel kinematic manipulator, shown in Fig. 2.

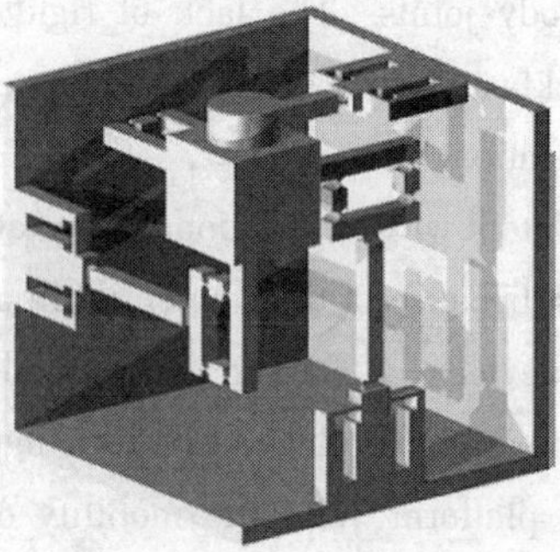

Fig. 1 Structural of 3-PRC compliant parallel mechanism

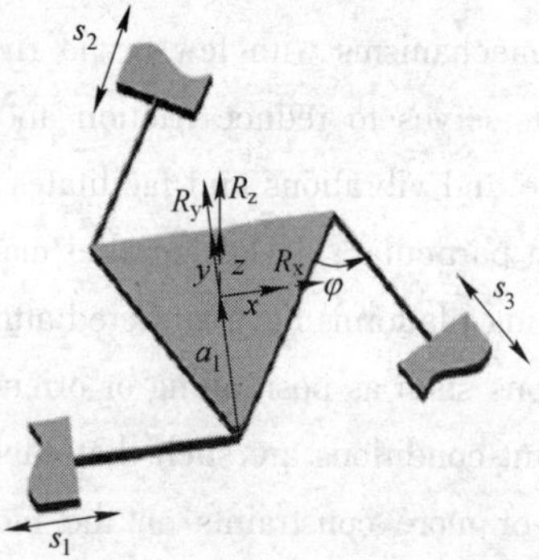

Fig. 2 Planar three-DoF parallel manipulator

Most existing micromanipulators can only provide 3-DoF of planar motion, or 3-DoF of combined spatial translation and rotation. Moreover, the method to replace rigid joints with flexure hinges makes it difficult to enhance the stiffness of the compliant parallel manipulator.

In this paper, the geometric constraint conditions used to enumerate the feasible limb structures of a class of 3-DoF rotational, fully compliant parallel manipulators are derived using the theory of reciprocal screws. A table of feasible limbs that can be used to construct 3-DoF rotational, fully spatial compliant parallel manipulators was developed based on the topology optimization theory for these geometric constraint conditions. When compared to conventional compliant parallel manipulators, the simulation results show the validity of the method for designing a fully spatial compliant parallel manipulator with high modal stiffness and vibration suppression abilities.

3.3.2 Geometry Constraint Conditions of a Conventional Parallel Manipulator

3.3.2.1 Screw and reciprocal screw theory

A unit screw is defined by a straight line with an associated pitch, and is represented by six screw coordinates[12]:

$$\$ = \begin{bmatrix} \boldsymbol{s} \\ \boldsymbol{r} \times \boldsymbol{s} + \lambda \boldsymbol{s} \end{bmatrix} \tag{1}$$

where $\boldsymbol{s}$ is a unit vector pointing in the direction of the screw axis; $\boldsymbol{r}$ is the position vector of any point on the screw axis with respect to a reference frame; λ is the pitch of the screw. If the pitch is equal to zero, the screw reduces to:

$$\$ = \begin{bmatrix} \boldsymbol{s} \\ \boldsymbol{r} \times \boldsymbol{s} \end{bmatrix} \tag{2}$$

An infinite-pitch screw is defined as:

$$\$ = \begin{bmatrix} 0 \\ \boldsymbol{s} \end{bmatrix} \tag{3}$$

A screw of intensity ρ is written as $\$ = \rho \hat{\$}$. A screw is defined as a twist if it represents the instantaneous motion of a rigid body, and is defined as a wrench if it represents a system of forces and moments acting on a rigid body.

For a serial manipulator, we may consider the motion of the end-effector to be twisted instantaneously about the joint axes of the open-loop chain. These instantaneous twists can be added linearly to give the instantaneous twist of the end-effector:

$$\$ = \begin{bmatrix} w \\ \vdots \\ v_0 \end{bmatrix} = [\hat{\$}_1, \hat{\$}_2, \cdots, \hat{\$}_n] \begin{bmatrix} \dot{q}_1 \\ \dot{q}_2 \\ \vdots \\ \dot{q}_n \end{bmatrix} \tag{4}$$

where $\$$ denotes the end-effector twist; w is the end-effector angular velocity; v_0 is the linear velocity of a point on the end-effector that is instantaneously coincident with the origin of a reference frame; $\dot{q}_j$ is the joint velocity rate.

Two screws, $\$$ and $\$_r$, are considered to be reciprocal to of each other if they satisfy the condition:

$$\$^{\mathrm{T}} \circ \$_r = 0 \tag{5}$$

where $\$^{\mathrm{T}} = [s_4 \quad s_5 \quad s_6 \quad s_1 \quad s_2 \quad s_3]$ is defined in such a way that $\$^{\mathrm{T}} \circ \$_r = s_4 s_{r1} + s_5 s_{r2} + s_6 s_{r3} + s_1 s_{r4} + s_2 s_{r5} + s_3 s_{r6}$, and denotes the component of the screw $\$$.

The twists associated with all the joints of a n-DoF serial chain form a n th order screw system, called a n-system, provided that all the joint screws are linearly independent. For spatial manipulators, if $n = 6$, there exists no screw that is reciprocal to the n-system. If $n < 6$, there exist n-6 linearly independent reciprocal screws that form a $(6 - n)$ system that is reciprocal to the n-system of twists.

3.3.2.2 Geometry constraint conditions

For a moving platform of a 3-DoF parallel manipulator to possess spherical motion, the twist system of the moving platform must be a 3-system of zero-pitch screws intersecting at the center of rotation. The reciprocal screws also form a 3-system of zero-pitch screws intersecting at the same point. Hence, the main purpose of the geometry constraint conditions is the enumeration of the limb structures that can provide three linearly independent zero-pitch constraint wrenches on the moving platform[12].

3.3.3 Topology of Optimization with Geometry Constraint Conditions

3.3.3.1 Structure synthesis with geometry constraint conditions

Geometry constraint conditions were analyzed, as shown in Fig. 3. Structure synthesis with geometry constraint conditions of a 3-DoF rotational fully spatial compliant parallel manipulator can be designed by the following

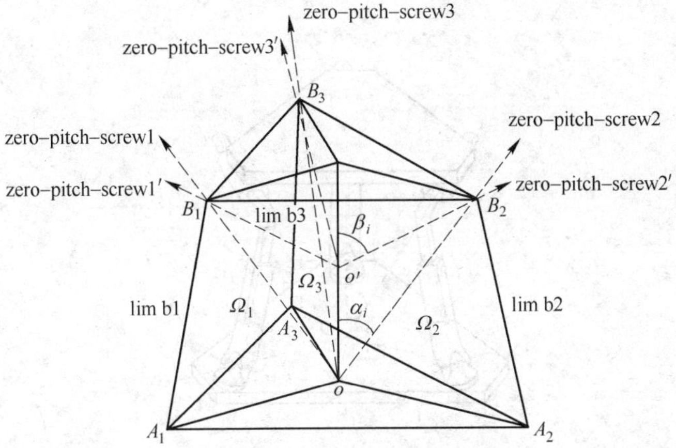

Fig. 3 Geometry constraint conditions of a 3-DoF rotational parallel manipulator

steps, as shown in Fig. 4.

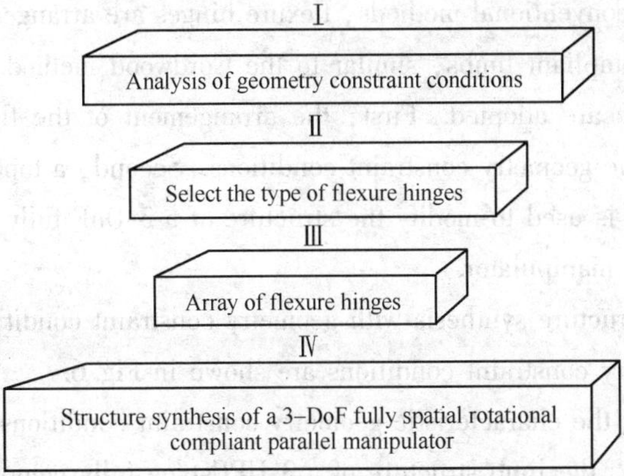

Fig. 4 Design process for fully spatial rotational compliant parallel manipulators

According to the literature[11], since there is a maximum of three linearly independent concurrent zero pitch wrenches, the limb wrench system may be a 1-, 2-or 3-system of zero pitch wrenches. The typical structure of a 3-

DoF rotational parallel manipulator is shown in Fig. 5.

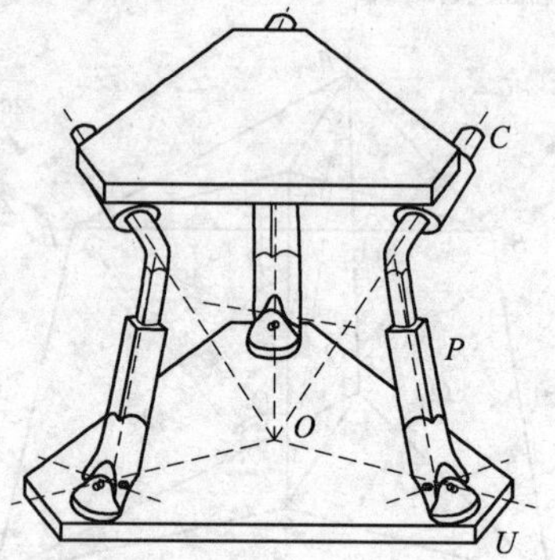

Fig. 5 A 3-UPC type rotational parallel manipulator

3.3.3.2 Structure of a 3-DoF fully spatial compliant parallel manipulator with an arrangement of flexure hinges in one plate

According to conventional methods, flexure hinges are arranged one by one to compose compliant limbs, similar to the Cordwood method. In this paper, two steps are adopted. First, the arrangement of the flexure hinges must satisfy the geometry constraint conditions. Second, a topology optimization method is used to modify the structure of a 3-DoF fully spatial compliant parallel manipulator.

Step 1 Structure synthesis with geometry constraint conditions

The geometry constraint conditions are shown in Fig. 6.

Considering the characteristic geometry constraint conditions provided by flexure hinges, the limb structure of a 3-UPC-type fully compliant parallel manipulator design is shown in Fig. 7.

Step 2 Topology optimization with flexure hinges

The topology optimization design for a compliant parallel manipulator considers stiffness and compliance as the primary functions. When these conditions and the flexibility of the material are limited, the stiffness of the material is maximized through a material distribution design. The optimiza-

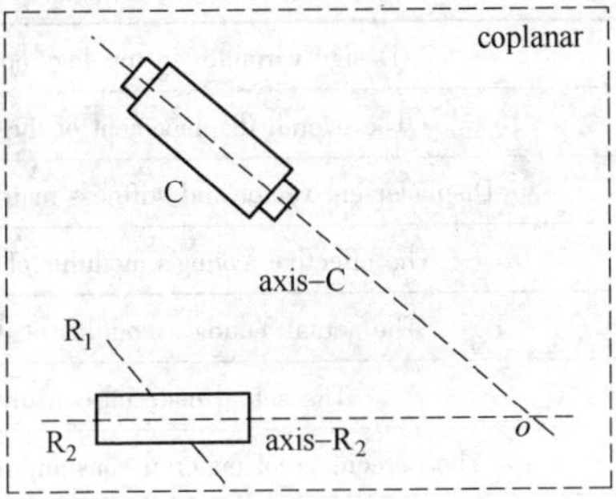

Fig. 6 Geometry constraint conditions of a 3-UPC type parallel manipulator

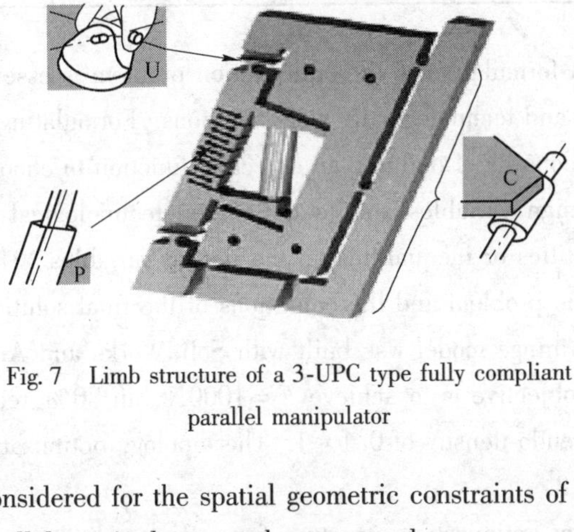

Fig. 7 Limb structure of a 3-UPC type fully compliant parallel manipulator

tion model considered for the spatial geometric constraints of a fully spatial compliant parallel manipulator can be expressed as:

$$F(\rho_i) = \boldsymbol{U}^{\mathrm{T}}\boldsymbol{K}\boldsymbol{U} = \sum_{i=1}^{n}\rho_i\boldsymbol{u}_i^{\mathrm{T}}\boldsymbol{k}_i\boldsymbol{u}_i, \quad i = 1,2,\cdots,n \tag{6}$$

Eq. (6) is minimized based on the following conditions:

$$E_i = E_0\rho_i,\ 0 \leqslant \rho_i \leqslant 1,\ i = 1,2,\cdots,n,\ F = KU,$$

$$V = fV^* = \sum_{i=1}^{n}\rho_i V_e,\ \$_{11} = \lambda_1\$_{13} = \lambda_2\$_{14} \tag{7}$$

The parameters for these equations are depicted in Table 1.

Table 1 Parameters depicted for topology optimization

ρ_i	Design variables in topology optimization
$\boldsymbol{U}$	The overall displacement of the institution
$\boldsymbol{u}_i$, $\boldsymbol{k}_i$	Displacement vector and stiffness matrix of the element
E_i	The effective Young's modulus of the element
E_0	The actual Young's modulus of the material
V	The actual material consumption
f	The percentage of material consumption which be set
V^*	The total design domain materials
V_e	Unit volume of material
$\$_{11}$, $\$_{13}$, $\$_{14}$	Axis vector of movement screw
λ_1, λ_2	Coefficient of $\$_{13}$ and $\$_{14}$, respectively

The correct formulation of the optimization problem is essential in achieving effective and technologically valid solutions. Formulating the optimization problem consists of defining an objective function to choose a set of independent design variables and to define the technological limitations in terms of equalities or inequalities on the design variables. The formulation depends on the problem and the conditions of the final solution.

The flexure hinge model was built with SolidWorks and Ansys software. The function objective is to achieve $F = 1000\text{N}$, an 80% reduction in the area and a pseudo-density of 0.1 ~ 1. The topology optimization process is shown in Fig. 8.

The topology optimization structure for a flexure limb of a 3-UPC-type compliant parallel manipulator is shown in Fig. 9.

3.3.4 Stiffness of a Fully Compliant Parallel Manipulator

3.3.4.1 Dynamic model of a 3-UPC type fully compliant parallel manipulator

Using finite element theory, we select an arbitrary cell and divide it into 8 nodes. The coordinate system is shown in Fig. 10.

The coordinate system $\{o_e - x_e y_e z_e\}$ is relative and fixed in the arbitrary

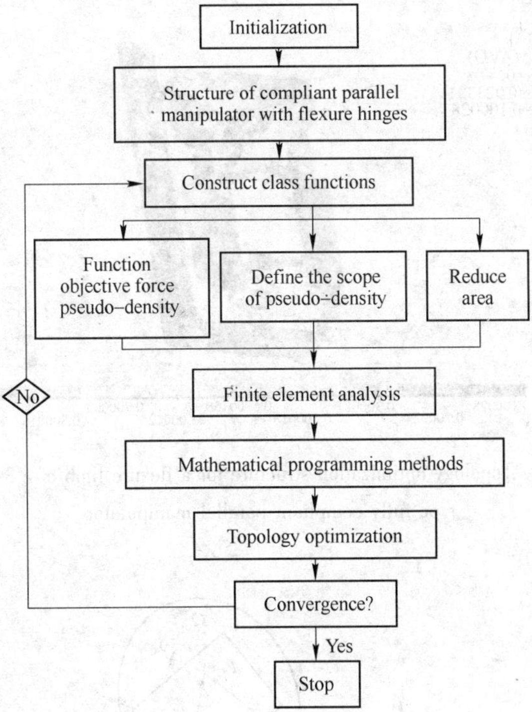

Fig. 8 Topology optimization algorithm chart

cell. The coordinate system $\{o - xyz\}$ is the reference frame. The variable T_e is the rotating coordinate transformation matrix to go between the $\{o_e - x_e y_e z_e\}$ and the $\{o - xyz\}$ reference frames. The variable p is an arbitrary point on the cell, which can be defined by the vector $\boldsymbol{l}_0$ in the reference frame. The variable p_d is a corresponding point of p and can be defined by the vector $\boldsymbol{l}$ in the reference frame. If we assume that the elastic displacement from p_d to p is $\boldsymbol{r}_e$, then:

$$\boldsymbol{l} = \boldsymbol{l}_0 + \boldsymbol{T}_e \boldsymbol{r}_e \tag{8}$$

The derivation of Eq. (8) is as follows:

$$\dot{\boldsymbol{l}} = \boldsymbol{T}_e \dot{\boldsymbol{r}}_e \tag{9}$$

From finite element theory, it is shown that:

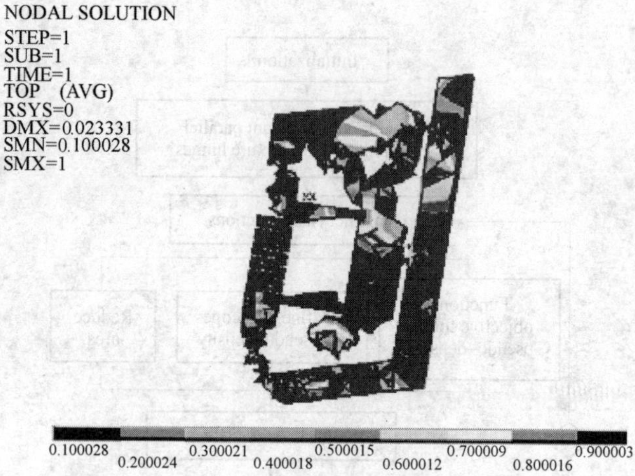

Fig. 9 Topology optimization structure for a flexure limb of a 3-UPC type fully compliant parallel manipulator

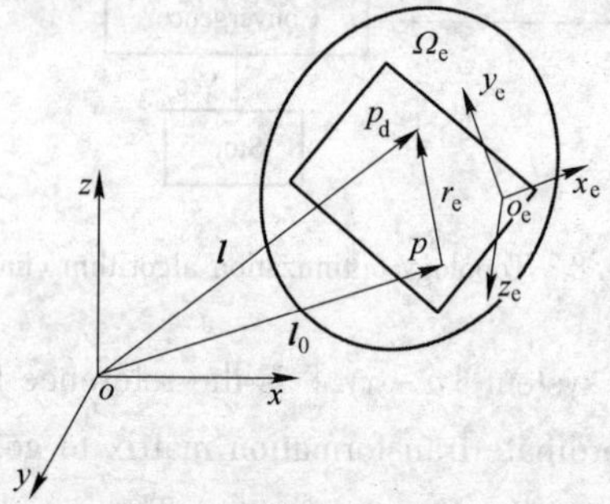

Fig. 10 Coordinate system of an arbitrary cell

$$\boldsymbol{r}_e = \boldsymbol{N}\boldsymbol{\delta}_e \tag{10}$$

where $\boldsymbol{N}$ and $\boldsymbol{\delta}_e$ are the shape function and the element node coordinate array in the relative coordinate system $\{o_e - x_e y_e z_e\}$, respectively.

Substituting Eq. (10) into Eq. (9) yields:

$$\dot{\boldsymbol{l}} = \boldsymbol{T}_e \boldsymbol{N}\boldsymbol{\delta}_e \tag{11}$$

The element node kinetic energy is defined as follows:

$$E_e = \frac{1}{2}\int_{l_e}\int_o\int \dot{l}^{\mathrm{T}}\dot{l}\mathrm{d}l_e = \frac{1}{2}\int_{l_e}\int_o\int \dot{\delta}_e^{\mathrm{T}}N^{\mathrm{T}}T_e^{\mathrm{T}}T_eN\dot{\delta}_e\mathrm{d}l_e \tag{12}$$

The total potential energy of the element node based on finite element theory is defined as:

$$P_e = \frac{1}{2}\int_{l_e}\int_o\int \boldsymbol{\delta}_e^{\mathrm{T}}\boldsymbol{B}_e^{\mathrm{T}}\boldsymbol{D}\boldsymbol{B}_e\boldsymbol{\delta}_e\mathrm{d}l_e - \int_{l_e}\int_o\int r_e^{\mathrm{T}}X_e\mathrm{d}l_e - \int_{l_e}\int \boldsymbol{r}_e^{\mathrm{T}}\overline{X}_e\mathrm{d}l_e \tag{13}$$

where $\boldsymbol{B}_e$ is element strain matrix; $\boldsymbol{D}$ is the elastic matrix; X_e is the volume force and $\overline{X}_e$ is the surface force applied on the element node.

Combining Eq. (12) with Eq. (13) provides the kinetic equation of a limb of a 3-UPC-type fully compliant parallel manipulator as:

$$\boldsymbol{M}\ddot{\boldsymbol{l}} + \boldsymbol{C}\dot{\boldsymbol{l}} + \boldsymbol{K}\boldsymbol{l} = \boldsymbol{Q}f_a \tag{14}$$

where $\boldsymbol{M}$ is the mass matrix; $\boldsymbol{C}$ is the damping matrix; $\boldsymbol{K}$ is the stiffness matrix; $\boldsymbol{Q}$ is generalized force matrix.

3.3.4.2 Natural frequency equation of a 3-UPC type fully compliant parallel manipulator

We assumed that the effects from damping can be ignored during the analysis of the natural frequency of a 3-UPC-type fully compliant parallel manipulator. Therefore, Eq. (14) can be rewritten as:

$$\boldsymbol{M}\ddot{\boldsymbol{l}} + \boldsymbol{K}\boldsymbol{l} = \boldsymbol{Q}_{l_{di}\to\Delta l_i}f_a \tag{15}$$

where $f_a = \mathrm{diag}(k_1, k_2, k_3)[-\Delta l_1, -\Delta l_2, -\Delta l_3]^{\mathrm{T}}$ and Δl_1 $(i=1\sim3)$ are the displacements of a piezoelectric actuator, k_i $(i=1\sim3)$ are the equivalent stiffness of the sub-chain, and $k_i = \boldsymbol{N}_i\boldsymbol{\delta}_e$. The displacements of the piezoelectric actuator in the different axes are defined as:

$$[-\Delta l_1, -\Delta l_1, -\Delta l_1]^{\mathrm{T}} = \sum_{j=1}^{m}\boldsymbol{J}_{l_{di}\to\Delta l_i}l_{\mathrm{d}i} \tag{16}$$

where $\boldsymbol{J}_{l_{di}\to\Delta l_i}$ is the transformation matrix from the element node to the sub-chain.

The second order differential equation describing the kinematics of the sub-chain can be written as:

$$M\ddot{l}+Q'l=0 \tag{17}$$

where $Q' = K + Q_{l_{di}\to\Delta l_i}\mathrm{diag}(k_1, k_2, k_3)\sum_{j=1}^{m}J_{l_{di}\to\Delta l_i}$

The equation for the natural frequency of a 3-UPC type fully compliant parallel manipulator can be derived as:

$$\|Q'-\omega^2 M\|=0 \tag{18}$$

3. 3. 5 Simulations

Two different steps are presented to demonstrate the design procedures for a compliant parallel manipulator. First, the flexibility characteristics of a 3-DoF UPC-type fully compliant parallel manipulator are demonstrated using Ansys software. Second, using the topology optimization method, the natural frequency is solved using Matlab.

Meshing for the Ansys simulation is shown in Fig. 11.

Fig. 11 Meshing of the sub-chain

The displacement of nodes in a piezoelectric ceramic actuator can be derived, as shown in Fig. 12. The partial node values for the displacement of the actuator are shown in Table 2.

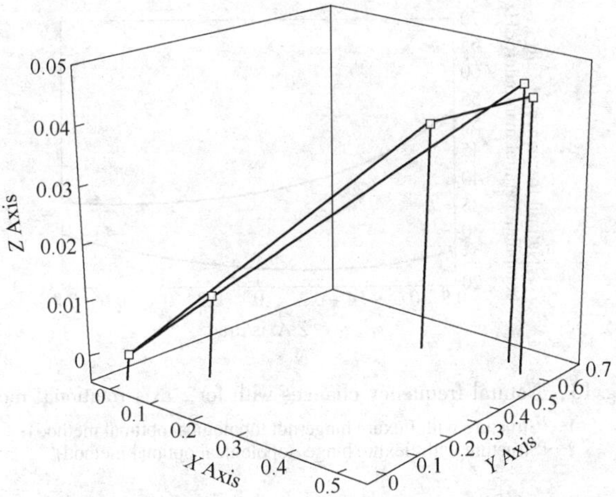

Fig. 12. Displacement of nodes in a piezoelectric ceramic actuator

Table 2 Partial node values of displacement for the piezoelectric actuator

Node	*UX*	*UY*	*UZ*	*USUM*
33	0. 36779E −03	0. 24149 E −04	0. 74575 E −06	0. 36859 E −03
96	0. 35128 E −03	−0. 15230 E −04	0. 70549 E −06	0. 35161 E −03
438	0. 31551 E −03	−0. 15897 E −04	0. 73552 E −06	0. 31591 E −03
942	0. 33263 E −03	−0. 16627 E −04	0. 65261 E −06	0. 33305 E −03
2609	0. 35741 E −03	0. 13098 E −04	0. 71980 E −06	0. 35765 E −03
2805	0. 33091 E −03	−0. 83106 E −05	0. 67533 E −06	0. 33101 E −03
4098	0. 29822 E −03	−0. 34338 E −04	0. 65413 E −06	0. 30019 E −03
4152	0. 30708 E −03	−0. 30550 E −04	0. 73839 E −06	0. 30860 E −03
10759	0. 30576 E −03	0. 47832 E −05	0. 66389 E −06	0. 30580 E −03
14354	0. 35789 E −03	0. 13438 E −04	0. 76433 E −06	0. 35814 E −03

Volume usage is set at 50% as a constraint, and a penalty factor of 3.0 is applied to eliminate grey phenomena during the optimization. The optimal result is shown in Fig. 9. The natural frequencies are shown in Fig. 13, Fig. 14 and Fig. 15 respectively.

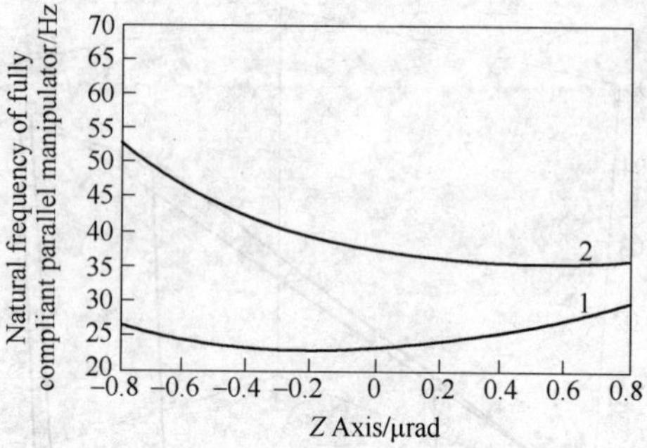

Fig. 13 Natural frequency changes with for z axis rotational motion

1— Structure with flexure hinge(not topological optimal method);
2— Structure with flexure hinge(topological optimal method)

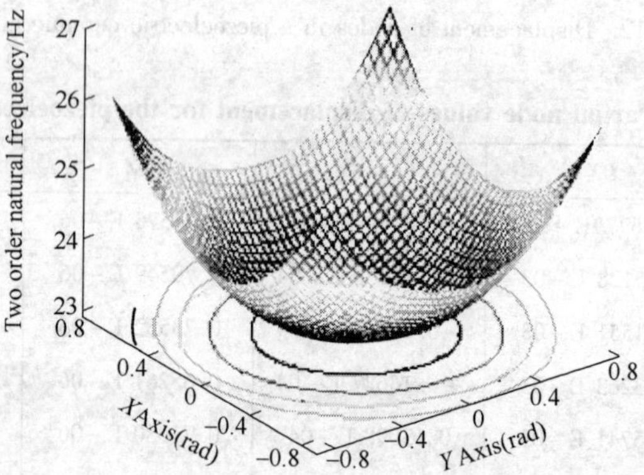

Fig. 14 Natural frequency changes for the X and Y axes of structural rotational motion of the flexure hinges (not using the topological optimization method)

3.3.6 Conclusions

This paper proposes a novel design method for a fully spatial compliant parallel manipulator. To overcome the disadvantages of the conventional design methods, simple rigid hinges were replaced with flexure hinges. The geometric constraint conditions are analyzed to reconfigure the sub-chain structure of the fully compliant parallel manipulator. As the method is adapted

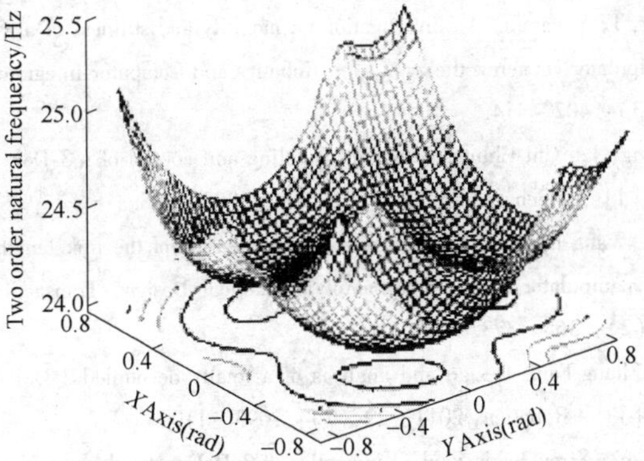

Fig. 15 Natural frequency changes for the X and Y axes of structural rotational motion of the flexure hinges (using the topological optimization method)

to the structure, the stiffness of the structure improves substantially. Furthermore, a topology optimization method was used to perfect the structure design process, and a 3-DoF UPC-type fully spatial compliant parallel manipulator was designed to illustrate the validity of the proposed method. The simulation results show that the proposed method is valid for designing a fully spatial compliant parallel manipulator.

References

[1] Howell L L. Compliant mechanisms [M]. New York, McGraw-Hill, 2001.

[2] Zhang X M. Topology optimization of compliant mechanisms [J]. Chinese Journal of Mechanical Engineering, 2003, 39: 47 ~ 51.

[3] Hao Qi, Wang Liping, Guan Liwen, Liu Xinjun. Dynamic analysis of a novel 3-PSP 3-DoF parallel manipulator [C] //Proceedings of the 2009 ASME/IFToMM International Conference on Reconfigurable Mechanisms and Robots, June22 ~ 24, London, United Kingdom, 2009: 309 ~ 314.

[4] Wang Jinsong, Wu Jun, Li Tiemin, Liu Xinjun. workspace and singularity analysis of a 3-DoF planar parallel manipulator with actuation redundancy [J]. Robotica, 2009, 27(1): 51 ~ 57.

[5] Li Yangmin, Xu Qingsong. Kinematic analysis and design of a new 3-DoF translational parallel manipulator [J]. Journal of Mechanical Design, Transactions of the ASME, 2006, 128

(4): 729 ~ 737.

[6] Xu Qingsong, Li Yangmin. An investigation on mobility and stiffness of a 3-DoF translational parallel manipulator via screw theory [J]. Robotics and Computer-Integrated Manufacturing, 2008, 24 (3): 402 ~ 414.

[7] Yen Ping-Lang, Lai Chi-Chung. Dynamic modeling and control of a 3-DoF Cartesian parallel manipulator [J]. Mechatronics, 2009, 19 (3): 390 ~ 398.

[8] Liu Xin-Jun, Wang Jinsong, Kim Jongwon. Determination of the link lengths for a spatial 3-DoF parallel manipulator [J]. Journal of Mechanical Design, Transaction of the ASME, 2006, 128 (2): 365 ~ 373.

[9] Zhang Dan, Zhang Fan. Design and analysis of a totally decoupled 3-DoF spherical parallel manipulator [J]. Robotica, 2011, 29 (7): 1093 ~ 1100.

[10] Sokolov Alexei, Xirouchakis Paul. Kinematics of 3-DoF parallel manipulator with an R-P-S joint structure [J]. Robotica, 2005, 23 (2): 207 ~ 217.

[11] Yuefa Fang, Lung-Wen Tsai. Structure synthesis of a class of 3-DoF rotational parallel manipulators [J]. IEEE Transactions on Robotics and Automation, 2004, 20 (1): 117 ~ 121.

[12] Hunt K. Kinematic Geometry of Mechanisms [M]. Cambridge, U. K: Oxford University Press, 1978.

[13] Ball R S. A Treatise on the Theory of Screws [M]. Cambridge, U. K: Cambridge University Press, 1990.

[14] Minoru Hashimoto, Yuicui Imamura. Design and characteristics of a parallel link compliant wrist [C] //Proceedings IEEE International Conference on Robotics and Automation, 1994, 3: 2457 ~ 2462.

[15] Byoung H K, John T W, Nicholas G D, et al. Analysis and design of parallel mechanisms with flexure joints [C] //Proceedings of the 2004 IEEE International Conference on Robotics & Automation, New Orieans, LA, 2004: 4097 ~ 4102.

[16] Yangmin Li, Qingsong Xu. Design and analysis of a new 3-DoF compliant parallel positioning platform for nanomanipulation [C] //Proceedings of 2005 5th IEEE Conference on Nanotechnology, Nagoya, Japan, 2005: 861 ~ 864.

[17] Qingsong Xu, Yangmin Li. Design modification of a 3-PRC compliant parallel micromanipulator for Micro/Nano scale manipulation [C] //Proceedings of the 7th IEEE International Conference on Nanotechnology, August 2 ~ 5, Hong Kong, 2007: 426 ~ 431.

[18] Jiming Huang, Yangmin Li, Xinhua Zhao. Optimization of a completely decoupled flexure-based parallel XY micro-motion stage [C] // Proceedings of the 2011 IEEE International Conference on Mechatronics and Automation, August 7 ~ 10, Beijing, China, 2011: 69 ~ 74.

[19] Jiming Huang, Yangmin Li. Analysis of a novel 2-DoF flexure hinge-based parallel micromanipulator in a polar coordinate system [C] //Proceedings of the 2010 IEEE International

Conference on Automation and Logistics, August 16 ~ 20, Hong Kong and Macau, 2010: 323 ~ 328.

[20] Boudewijn R, Dannis M B, Meint J, et al. Design and fabrication of a planar three-DoFs MEMS-based manipulator [J]. Journal of Microelectromechanical Systems, 2010: 19 (5): 1116 ~ 1130.

4 全柔顺并联机构微分运动及振动抑制研究

4.1 Vibration Active Suppress of a 4-DoF Fully Compliant Parallel Manipulator Based on Discrete Time Sliding Mode Control

An vibration active suppress control approach of a 4-DoF fully compliant parallel manipulator with piezoelectric driver and strain gauges is presented by using discrete sliding mode control. The differential equation method and modal truncation technique are applied to obtain the kinematic/dynamic model of this system. The stiffness matrix is obtained according to vibration suppress and control requirements. The discrete sliding mode control method with disturbance observer is applied to design the vibration active suppress controller. Kalman filter is also employed to construct state estimator. Natural frequencies of four flexure limbs are obtained by using stiffness matrix and experimental modal testing. The experimental results shown that the proposed controller can effectively reduce the vibration response of the 4-DoF fully compliant parallel manipulator.

4.1.1 Introduction

Parallel manipulators have attracted a significant amount of attention among researchers and engineers in the past two decades because they possess inherent advantages over conventional serial manipulators, such as high rigidity, load capacity, velocity, and precision. However, as for any mechanical system composed of conventional joints, traditional parallel manipulators suffer errors due to backlash, hysteresis, and manufacturing flaws in the joints. Hence, it is a major challenge to achieve ultra-high precision using conventional rigid joints. As micro- and nanotechnology piezoelectric actuating techniques become some of the most important techniques in the fields of micromanipulation and microposition, compliant mechanisms have attracted more attention. Compliant mechanisms are flexible structures that generate some desired motions through elastic deformation instead of through rigid linkages, as in rigid body mechanisms. Compliant mechanisms are suitable for use in the fields of precision engineering, as a result of the monolithic nature of the compliant mechanisms with few or no rigid joints. The lack of rigid joints serves to reduce friction and backlash losses, part assembly costs, noise and vibrations and facilitates unitization[1,2].

The use of adaptable components or smart structures[3], the integration of actuators, sensors, and control directly with a material or structure, opens up interesting new possibilities for modern machine tools. These machine tools, for example, parallel kinematic machines[4], are built for high processing speeds and accuracy and use new technologies such as lightweight structures or linear direct drives[5,6]. Due to the lightweight design, vibrations and deformations of the machine tool structure become an important issue affecting the machine performance negatively. Design measures along are often not sufficient anymore in order to achieve a desired static and dynamic stiffness, which are the major limiting factors for machining quality, especially if high processing speeds, as well as high accuracies, are required. Two active vibration control concepts were designed for an adaptronic component of a parallel kinematic machine tool which is modeled as a flexible multibody system model including a nonlinear flatness-based position control[7]. An integrated inversion-based approach is presented in reference [8] to compensate for all adverse affects-creep, hysteresis and vibrations. A virtual spring is integrated with the structure flexibility to establish an internal model for the external disturbance and low-order control algorithm was developed to regulate an infinitely dimensional flexible beam subject to non-decaying, harmonic disturbances of known frequency in reference [9]. Dynamic stiffness matrix and vibration analysis are considered in reference [10 ~ 12] respectively.

As flexure joints are frequently used in precision motion stages and micro-robotic mechanisms due to their monolithic construction. Based on the differentiation of kinematic equations, the error transformation matrix was derived in reference [13]. To achieve a minimum deflection in solution of inverse kinematics problem, the differential evolution method was used and the static stiffness of manipulator was discussed in reference [14].

As a popular non-linear control method, sliding model control (SMC) or variable structure control has been considered to solve the non-linear motion existing in parallel manipulator. According to all of the parameter uncertainties, a state-feedback control with variable structure control was proposed for a two-link flexible-joint robot[15]. A neural-adaptive sliding mode control for the tracking control of 4-SPS (PS) type parallel manipulator was presented by reference [16]. Integral sliding mode controllers were designed in joint space and in task space are compared using MATLAB simulations for a 6-DoF parallel robot in reference [17].

In this paper, the vibration control of 4-DoF fully compliant parallel manipulator is investigated by using discrete-time sliding mode controller with equivalent law and experimental modal test and experimental studies are undertaken to verify the presented method. Firstly, differential dynamic model of 4-DoF fully compliant parallel manipulator with piezoelectric actuations and strain gauge sensors were obtained. Secondly, a discrete-time sliding mode controller with Kalman filter is proposed. Finally, experimental mode test of the 4-DoF fully compliant parallel manipulator is performed to obtain its modal parameters, and the experimental simulations are carried out.

4.1.2 Dynamic Model of 4-DoF Compliant Parallel Manipulator

4-DoF $R_1PR_2R_3R_4$ type fully compliant parallel manipulator using flexure hinges at one planar consists of a moving platform, a fixed base, and four limbs with identical kinematic structure. Each limb connects the moving platform to the fixed base by one compliant limb which composed by one flexure rotational hinge (R_1), one flexure prismatic hinge (P) and followed by three flexure rotational hinges ($R_2R_3R_4$) in sequence, where the *P* joint is fixed between two group rotational joints and actuated by a piezoelectric (PZT) actuator. Geometrical condition of two group rotational joints is that the axis of R_1 parallel with R_2 and the axis of R_3 intersect with R_4 at the point o''. Structure of 4-DoF $R_1PR_2R_3R_4$ type fully compliant parallel manipulator is shown as Fig. 1.

Configure of fully hinges is a composition shaft configuration with very high torsional stiffness. PZT actuator offer the values involving smooth motion, high accuracy, etc., which make them much suitable for active vibration isolation for micro-elcetro-mechanical systems (MEMS).

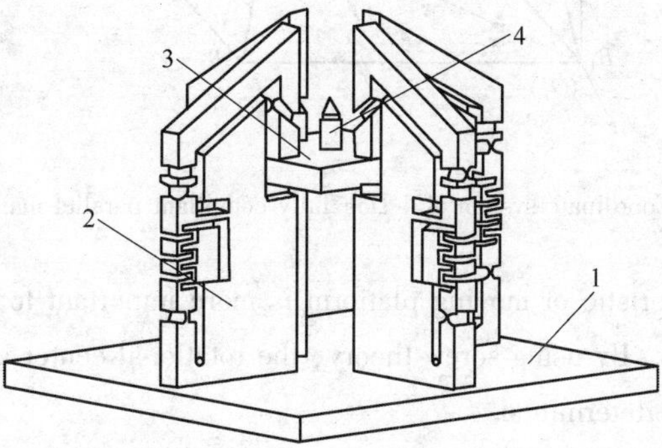

Fig. 1 Structure of 4-DoF fully compliant parallel manipulator

1—fixed base; 2—flexure limb; 3—moving platform; 4—positioner

As shown in Fig. 2, let the reference coordinate system located at the center of fixed base, with axis directions determined by orthogonal set of unit vectors, defined as $\{o_b - xyz\}$. Fix a coordinate system $\{o_a - uw\boldsymbol{v}\}$ to the moving platform at the center o_a. One of relative coordinate system $\{o_i' - x_i'y_i'z_i'\}$ is fixed at the arbitrary point B_i ($i = 1 \sim 4$) with vector $\overrightarrow{o_ix_i'}$ toward the center of fixed platform $\boldsymbol{o_b}$. The other relative coordinate system $\{o'' - x''y''z''\}$ is fixed at the intersect point $\boldsymbol{o''}$ with vector $\overrightarrow{\boldsymbol{o''z''}}$ along with direction of $\overrightarrow{o_bo_a}$ and vector $\overrightarrow{o''x''}$ along with direction of $\overrightarrow{B_io_b}$. Due to symmetric structure, the initial length of each limb is l_i and $o_aA_i = a_i o_bB_i = b_i$ $(i = 1 \sim 4)$.

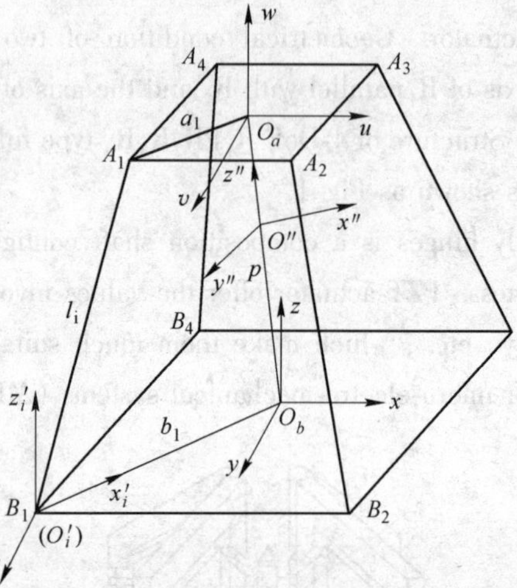

Fig. 2 Coordinate system of 4-DoF fully compliant parallel manipulator

The characteristic of moving platform is more important for kinematic analysis process. By using screw theory, the rotational center and motion direction can be determined.

The rotational center is located at point o'' and motion direction along with vector $\overrightarrow{o''z''}$.

Proof: Limb structure is shown as Fig. 1, and $\$_i$ is denoted as screw of $i^{\#}$ joint, and the screw style of rotational joint is $\$ = \begin{bmatrix} s \\ r \times s \end{bmatrix}$, the screw style of prismatic joint is $\$ = \begin{bmatrix} 0 \\ s \end{bmatrix}$ (where s is a unit vector pointing at the direction of the screw axis, and r is position vector of any point on the joint axis respect to a reference frame). Assumed that the coordinate of intersect point is (x, y, z), and the screw axis direction of R_4 and R_5 are set with $s_4 = (l_1 \quad m_1 \quad n_1)$ and $s_5 = (l_2 \quad m_2 \quad n_2)$ respectively. So the screw system of limb is shown as follows:

$$\begin{cases} \$_1 = (1 \quad 0 \quad 0 \quad 0 \quad 0 \quad 0) \\ \$_2 = (0 \quad 0 \quad 0 \quad 0 \quad \sin\alpha \quad \cos\alpha) \\ \$_3 = (1 \quad 0 \quad 0 \quad -\sin\alpha \quad 0 \quad \cos\alpha) \\ \$_4 = (l_1 - x \quad m_1 - y \quad n_1 - z \quad m_1x - l_1y \quad n_1y - m_1z \quad l_1z - n_1x) \\ \$_5 = (l_2 - x \quad m_2 - y \quad n_2 - z \quad m_2x - l_2y \quad n_2y - m_2z \quad l_2z - n_2x) \end{cases} \tag{1}$$

Reciprocal screw (restraint condition) is derived as follows:

$$\hat{\$} = (1 \quad 0 \quad 0 \quad 0 \quad M \quad N) \tag{2}$$

where

$$M = \frac{n_2m_1x + m_2xz + l_2yz + l_2n_1y - m_1xz - n_1m_2x - l_1n_2y - l_2yz}{m_2z - m_1z + n_1y - m_2n_1 - n_2y + m_1n_2}$$

$$N = \frac{(y - m_2)(xm_1 - yl_1) - (y - m_1)(xm_2 - yl_2)}{(y - m_1)(z - n_2) - (y - m_2)(z - n_1)}$$

Based on concept of screw and reciprocal screw, $\hat{\$}$ denotes a pure force restraint acting on the moving platform. Assumed that the coordinate of acting point is (x', y', z'), and then we can get $\hat{\$}' = (1 \quad 0 \quad 0 \quad -y' \quad 0 \quad z')$. Compared with two equations $\hat{\$}$ and $\hat{\$}'$, the parameters can be derived as follows:

$$y' = 0, \ M = 0, \ z' = N \tag{3}$$

Solve with Eq. (2), we can obtain:

$$x' = \text{arbitrary}, \ y' = 0, \ z' = z \tag{4}$$

It illuminated that the direction of restraint force along with axis of x and acting on the intersect point o''.

4.1.3 Differential Kinematic Model of 4-DoF Compliant Parallel Manipulator

4.1.3.1 Kinematic of 4-DoF fully compliant parallel manipulator

Let δ_x, δ_y, δ_z represent rotational angles of moving platform. It is assumed that the small angel approximations hold by PZT actuators and then the rotational matrics among the coordinate system $\{o'' - x''y''z''\}$ for $i^{\#}$ flexure limb are as follows:

$$R = R_{o''}(x'',\delta_x)R_{o''}(y'',\delta_y)R_{o''}(z'',\delta_z) = \begin{bmatrix} 1 & -\delta_z & \delta_y \\ \delta_z & 1 & -\delta_x \\ -\delta_y & \delta_x & 1 \end{bmatrix} \tag{5}$$

The expressions for the displacement of the reference frame of each upper arm with respect to initial reference frame of upper arm are:

$$\hat{\boldsymbol{A}}_i^{o_b} = \begin{bmatrix} \hat{\boldsymbol{A}}_{ix}^{o_b} \\ \hat{\boldsymbol{A}}_{iy}^{o_b} \\ \hat{\boldsymbol{A}}_{iz}^{o_b} \\ 1 \end{bmatrix} = \boldsymbol{c} + \boldsymbol{R}\boldsymbol{A}_i^{o_b} = \begin{bmatrix} 1 & -\delta_z & \delta_y & 0 \\ \delta_z & 1 & -\delta_x & 0 \\ -\delta_y & \delta_x & 1 & \Delta h \\ 0 & 0 & 0 & 1 \end{bmatrix} \begin{bmatrix} \boldsymbol{A}_{ix}^{o_b} \\ \boldsymbol{A}_{iy}^{o_b} \\ \boldsymbol{A}_{iz}^{o_b} \\ 1 \end{bmatrix} \tag{6}$$

Differentiating Eq. (6) with respect to time:

$$\boldsymbol{v}_{ai} = \dot{\boldsymbol{c}} + \dot{\boldsymbol{R}}\boldsymbol{A}_i^{o_b} = \begin{bmatrix} 0 & -\omega_z & \omega_y & 0 \\ \omega_z & 0 & -\omega_x & 0 \\ -\omega_y & \omega_x & 0 & v_z \\ 0 & 0 & 0 & 1 \end{bmatrix} \begin{bmatrix} \boldsymbol{A}_{ix}^{o_b} \\ \boldsymbol{A}_{iy}^{o_b} \\ \boldsymbol{A}_{iz}^{o_b} \\ 1 \end{bmatrix} = \boldsymbol{J}_{ai}^{o_b}\dot{\boldsymbol{x}} \tag{7}$$

where $\dot{\boldsymbol{x}}$ is generalized differential velocity of 4-DoF $R_1PR_2R_3R_4$ type fully compliant parallel manipulator, and $\boldsymbol{J}_{ai}^{o_b}$ is velocity Jacobian matrix.

Differentiating Eq. (7) with respect to time, we can obtain:

$$\dot{\boldsymbol{v}}_{ai} = \boldsymbol{a}_{ai} = \ddot{\boldsymbol{c}} + \dot{\boldsymbol{\omega}} \times \boldsymbol{R}\boldsymbol{A}_i = \boldsymbol{J}_{ai}\ddot{\boldsymbol{x}} + \dot{\boldsymbol{J}}_{ai}\dot{\boldsymbol{x}} \tag{8}$$

where $\ddot{\boldsymbol{x}}$ is generalized differential acceleration of 4-DoF $R_1PR_2R_3R_4$ type fully compliant parallel manipulator.

4.1.3.2 Kinematic of PZT actuator

Vector of PZT is denoted as follows:

$$\boldsymbol{l}_i = \boldsymbol{p} + \boldsymbol{R}\boldsymbol{a}_i - \boldsymbol{b}_i \tag{9}$$

Velocity of PZT actuator is:

$$\dot{\boldsymbol{l}}_i = \boldsymbol{l}_{ni}^{\mathrm{T}}\boldsymbol{v}_{ai} \tag{10}$$

where $\boldsymbol{l}_{ni}$ is the unit direction vector of PZT, and $\boldsymbol{l}_{ni} = \boldsymbol{l}_i / |\boldsymbol{l}_i|$.

Submit Eq. (10) into Eq. (9), which can be assembled into a matrix form:

$$\boldsymbol{v}_1 = [L_n^{\mathrm{T}} \quad (RA_i \times L_n)^{\mathrm{T}}]\begin{bmatrix}\dot{p}\\ \omega\end{bmatrix} = J_1 \dot{x} \tag{11}$$

where $\boldsymbol{L}_n$ is matrix with unit direction vector of PZT, $\boldsymbol{v}_1$ is vector matrix of PZT and $\boldsymbol{v}_1 = [v_{1x} \quad v_{2x} \quad v_{3x} \quad v_{4x}]^{\mathrm{T}}$, $\boldsymbol{J}_1$ is Jacobian matrix of generalized velocity from moving platform to PZT actuators.

Assumed that angular velocity ω_{ai} rotate around the point B_i is:

$$\boldsymbol{\omega}_{ai} \times \boldsymbol{l}_i = \boldsymbol{v}_{ai} - \boldsymbol{l}_{ni}^{\mathrm{T}} \boldsymbol{v}_{ai} \boldsymbol{l}_{ni} \tag{12}$$

Cross-multiplying both sides of Eq. (12) by l_{ni}:

$$\omega_{ai} = \frac{\boldsymbol{l}_{ni}}{|\boldsymbol{l}_i|} \times \boldsymbol{v}_{ai} = \boldsymbol{J}_{\omega} \cdot \boldsymbol{v}_{ai} \tag{13}$$

where $\boldsymbol{J}_{\omega}$ is angular velocity matrix of PZT.

Differential Eq. (13) with respect to time, we can obtain the angular acceleration $\dot{\boldsymbol{\omega}}_{ai}$:

$$\dot{\boldsymbol{\omega}}_{ai} = \frac{\boldsymbol{l}_{ni} \times \dot{\boldsymbol{v}}_{ai} + \omega_{ai} \times \boldsymbol{l}_{ni} \times \boldsymbol{v}_{ai} - \omega_{ai} \dot{\boldsymbol{l}}_i}{|\boldsymbol{l}_i|} = \boldsymbol{J}_{\omega} \dot{\boldsymbol{v}}_{ai} + \dot{\boldsymbol{J}}_{\omega} \boldsymbol{v}_{ai} \tag{14}$$

4.1.3.3 Differential dynamics model of 4-DoF $R_1PR_2R_3R_4$ type fully compliant parallel manipulator

Let $\boldsymbol{F}_p$ and $\boldsymbol{M}_p$ represent the force and moment exerted by the ith upper joint on the moving platform at the ith flexure limb. Moving platform's contribution to the set of generalized active forces is:

$$\boldsymbol{F}_p = \sum_{i=1}^{4} \boldsymbol{f}_i \frac{\boldsymbol{l}_{ni}}{|\boldsymbol{l}_i|} + \boldsymbol{m}_{\mathrm{p}} \boldsymbol{g} \tag{15}$$

$$\boldsymbol{M}_{\mathrm{p}} = \sum_{i=1}^{4} \left(\boldsymbol{Ra}_i \times \boldsymbol{f}_i \frac{\boldsymbol{l}_{ni}}{|\boldsymbol{l}_i|} \right) \tag{16}$$

where $\boldsymbol{f}_i$ is PZT driving force of each flexure limb, and $\boldsymbol{m}_{\mathrm{p}}$ is the mass of moving platform.

Generalized active force can be derived as follows:

$$\boldsymbol{K}_{r1} = \boldsymbol{v}_{o_a}^{\mathrm{T}} \cdot \boldsymbol{F}_{\mathrm{p}} + \boldsymbol{\omega}_{o_a}^{\mathrm{T}} \cdot \boldsymbol{M}_{\mathrm{p}} \tag{17}$$

where $\boldsymbol{v}_{o_a}^{\mathrm{T}}$ is differential velocity matrix, and $\boldsymbol{v}_{o_a}^{\mathrm{T}} = [0 \quad 0 \quad \mathrm{d}z]^{\mathrm{T}}$. $\boldsymbol{\omega}_{o_a}^{\mathrm{T}}$ is differential angular velocity matrix, and $\boldsymbol{\omega}_{o_a}^{\mathrm{T}} = \begin{bmatrix} 0 & -\omega_z & \omega_y \\ \omega_z & 0 & -\omega_x \\ -\omega_y & \omega_x & 0 \end{bmatrix}$.

Generalized inertial force can be derived as follows:

$$\boldsymbol{K}_{r1}^* = -\boldsymbol{v}_{o_a}^{\mathrm{T}} \boldsymbol{m}_{\mathrm{p}} \ddot{\boldsymbol{c}} - \boldsymbol{\omega}_{o_a}^{\mathrm{T}} (\boldsymbol{I}_{\mathrm{p}} \dot{\boldsymbol{\omega}} + \boldsymbol{\omega} \times \boldsymbol{I}_{\mathrm{p}} \boldsymbol{\omega}) \tag{18}$$

where $\boldsymbol{I}_{\mathrm{p}} = \boldsymbol{R} \boldsymbol{I}_{o_a}^{o_b} \boldsymbol{R}^{\mathrm{T}}$, and $\boldsymbol{I}_{o_a}^{o_b}$ is moving platform inertial matrix reference to $\{o_b - xyz\}$ coordinate system.

The holonomic generalized active force of flexure limb gravity for the ith $(i = 1 \sim 4)$ generalized speed is:

$$\boldsymbol{K}_{di} = \boldsymbol{v}_{li}^{\mathrm{T}} \boldsymbol{m}_{di} \boldsymbol{g} \quad (i = 1 \sim 4) \tag{19}$$

Generalized active force of PZT actuator gravity can be obtained as follows:

$$\boldsymbol{K}_{ui} = \boldsymbol{v}_{ui}^{\mathrm{T}} \boldsymbol{m}_{ui} \boldsymbol{g} \quad (i = 1 \sim 4) \tag{20}$$

where $\boldsymbol{m}_{di}$ is the mass of $\boldsymbol{i}^{\mathrm{th}}$ flexure limb, and $\boldsymbol{m}_{ui}$ is the mass of $\boldsymbol{i}^{\mathrm{th}}$ PZT actuator, $\boldsymbol{v}_{li}$ and $\boldsymbol{v}_{ui}$ are the speed of the center of mass at flexure limb and PZT actuator respectively.

Generalized inertial force of flexure limb is:

$$\boldsymbol{K}_{di}^* = -\boldsymbol{v}_{li}^{\mathrm{T}} \boldsymbol{m}_{di} \boldsymbol{v}_{o_a}^{\mathrm{T}} - \boldsymbol{\omega}_{o_a}^{\mathrm{T}} (\boldsymbol{I}_d \dot{\boldsymbol{\omega}}_{ai} + \boldsymbol{\omega}_{ai} \times \boldsymbol{I}_d \boldsymbol{\omega}_{ai}) \tag{21}$$

Generalized inertial force of PZT actuator is:

$$\boldsymbol{K}_{ui}^* = -\boldsymbol{v}_{li}^{\mathrm{T}} \boldsymbol{m}_{di} \boldsymbol{v}_{o_a}^{\mathrm{T}} - \boldsymbol{\omega}_{o_a}^{\mathrm{T}} (\boldsymbol{I}_u \dot{\boldsymbol{\omega}}_{ai} + \boldsymbol{\omega}_{ai} \times \boldsymbol{I}_u \boldsymbol{\omega}_{ai}) \tag{22}$$

where $\boldsymbol{I}_d$ and $\boldsymbol{I}_u$ are inertial matrix reference to the $\{o_b - xyz\}$ coordinate system.

Kane's dynamics equation is:

$$\boldsymbol{K}_r + \boldsymbol{K}_r^* = 0 \tag{23}$$

where $\boldsymbol{K}_r$ is generalized active force of 4-DoF $\mathrm{R_1PR_2R_3R_4}$ type fully compliant parallel manipulator, and $\boldsymbol{K}_r = \boldsymbol{K}_{r1} + \sum_{i=1}^{4} (\boldsymbol{K}_{di} + \boldsymbol{K}_{ui})$, $\boldsymbol{K}_r^*$ is generalized inertial force of 4 - DoF $\mathrm{R_1PR_2R_3R_4}$ type fully compliant parallel manipulator, and $\boldsymbol{K}_r^* = \boldsymbol{K}_{r1}^* + \sum_{i=1}^{4} (\boldsymbol{K}_{di}^* + \boldsymbol{K}_{ui}^*)$.

The Eq. (23) can be arranged to the dynamic form:

$$M\ddot{x} + C\dot{x} + G = J_l^{\mathrm{T}} f_a \tag{24}$$

where M is the mass matrix of 4 - DoF $R_1PR_2R_3R_4$ type fully compliant parallel manipulator, and:

$$M = \begin{bmatrix} m_{\mathrm{p}} & 0 \\ 0 & RI_{o_a}^{o_b}R^{\mathrm{T}} \end{bmatrix} + \sum_{i=1}^{4} M_i$$

$$M_i = v_{li}^{\mathrm{T}} m_{di} v_{li} + v_{ui}^{\mathrm{T}} m_{ui} v_{ui} + \omega_{ai}^{\mathrm{T}} (I_d + I_u) \omega_{ai}$$

f_a is the matrix of driving force by PZT actuators, and $f_a = [f_1 \quad \cdots \quad f_4]^{\mathrm{T}}$, C is Coriolls force and G is the gravitational term.

4.1.3.4 Natural frequency equations

Without loss of generality, as this compliant parallel manipulator has differential motion characteristic, so the Coriolls force and gravitational term can be ignored. Eq. (24) can be simplified as follows:

$$M\ddot{x} = J_l^{\mathrm{T}} f_a \tag{25}$$

where $f_a = [f_1 \quad \cdots \quad f_4]^{\mathrm{T}} = \mathrm{diag}(k_1 \quad \cdots \quad k_4)[\Delta l_1 \quad \cdots \quad \Delta l_4]^{\mathrm{T}}$, k_i are equivalent stiffness of PZT actuator, and $k_i = \frac{\beta_e A}{h}$, β_e is piezoelectric coefficient, and A is contact area, h is the length of PZT actuator, $[\Delta l_1 \quad \cdots \quad \Delta l_4]^T = J_l x$.

The final dynamics equation of this compliant parallel manipulator system can be written as:

$$M\ddot{x} + Kx = 0 \tag{26}$$

where $K = J_1^{\mathrm{T}} \mathrm{diag}(k_1 \quad \cdots \quad k_4) J_l$.

The natural frequency equation of 4-DoF $R_1PR_2R_3R_4$ type fully compliant parallel manipulator can be obtained as follows:

$$\|K - \omega^2 M\| = 0 \tag{27}$$

where ω is the natural frequency.

4.1.4 Sliding Mode Controller

4.1.4.1 System dynamic modelling

It is assumed that the same magnitude but the opposite direction of electric field is applied to the upper and lower piezoelectric actuators so as to create a pure bending moment for vibration suppression. The moment is proportional to the control voltage. These bending moments induced by the actuators is given as follows:

$$T_A = T_B = -\beta_e E_p b_p (t_p + t_a) V_{\text{in}} \tag{28}$$

where β_e is piezoelectric coefficient, E_p, b_p, t_p are elastic module, width, and thickness of PZT actuators respectively, t_a is the thickness of the flexure limb element, V_{in} is the vector of input voltage to the piezoelectric actuators.

Strain is the amount of deformation of a body due to an applied force. Strain gauge is the most common sensor for measuring strain. A strain gauge's electrical resistance varies in proportion to the amount of strain placed on it. The deformation is mainly bending deformation for the flexure limbs. The strain ε in axial direction is given by:

$$\varepsilon = \varepsilon_L + \varepsilon_B \tag{29}$$

where ε_L is compression or tension strain and ε_B is longitudinal strain due to bending deformation, and they are given by:

$$\varepsilon_L = K_{di} \cdot (\Delta l_1 \quad \cdots \quad \Delta l_4)^{\mathrm{T}} \text{ and } \varepsilon_B = K_{ui} \cdot (\Delta l_1 \quad \cdots \quad \Delta l_4)^{\mathrm{T}}$$

Employing the previous modeling of the structure, actuator and sensor, the dynamic equations of the fully compliant parallel manipulator equipped with piezoelectric actuators and strain gauge transducer can be expressed as follows:

$$\begin{aligned} \boldsymbol{M}\ddot{\boldsymbol{x}} + \boldsymbol{C}\dot{\boldsymbol{x}} + \boldsymbol{G} &= \boldsymbol{J}_l^{\mathrm{T}} \boldsymbol{f}_a + \xi_a \boldsymbol{V}_{\text{in}} \\ \boldsymbol{y} &= \xi_s \boldsymbol{x} \end{aligned} \tag{30}$$

where ξ_a is the systematic control matrix related to configuration of actuators, and ξ_s is the systematic output matrix determined by configuration of sensors.

Taking into account measurement noise v, the state-space model for the fully compliant parallel manipulator can be rewritten as follows:

$$\begin{aligned} \boldsymbol{x}(k) &= \boldsymbol{A}\boldsymbol{x}(k-1) + \boldsymbol{B}(u(k) + w(k)) \\ \boldsymbol{y}_v(k) &= \boldsymbol{C}\boldsymbol{x}(k) + v(k) \end{aligned} \tag{31}$$

where $\boldsymbol{x}(n)$ is a discrete state, $\boldsymbol{A} \in \mathbf{R}^{4\times 4}$, $\boldsymbol{B}$, $\boldsymbol{C} \in \mathbf{R}^{4}$ are the system matrix and vectors of the discrete-time model with sampling period T, and d is a time delay in the discrete-time representation.

4.1.4.2 Design of sliding mode controller

The controller adopted in this research is a discrete-time sliding mode controller with disturbance observer. For the purpose of suppressing the vibration of the proposed fully compliant parallel manipulator, the controller design is essentially a regulator design. Since the modal positions and velocities of the system are not directly measurable, and a state observer is employed to obtain the estimated states. Block diagram of control system is shown as Fig. 3.

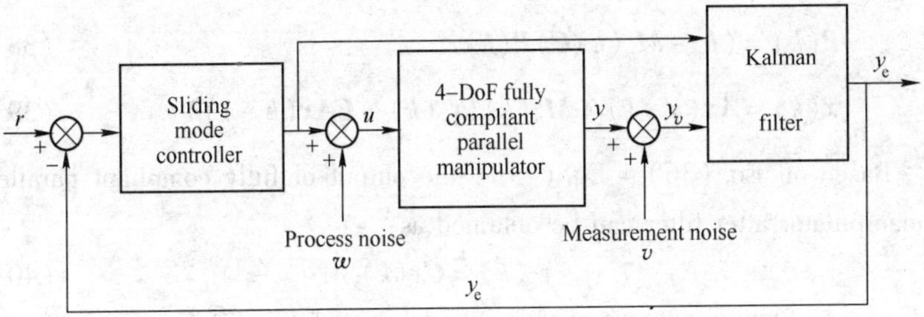

Fig. 3 Block diagram of the control system

Assumed that the position input is $r_i(k)$, and its rate of change is $\mathrm{d}r_i(k)$,

$$\boldsymbol{R} = [r_i(\boldsymbol{k});\mathrm{d}r_i(k)];\ \boldsymbol{R}_1 = [r_i(k+1);\mathrm{d}r_i(k+1)] \quad (i=1\sim 4)$$

So we can obtain:

$$r_i(k+1) = 2r_i(k) - r_i(k-1);\ \mathrm{d}r_i(k+1) = 2\mathrm{d}r_i(k) - \mathrm{d}r_i(k-1)$$

The switching function is:

$$s(k) = \boldsymbol{C}_e\boldsymbol{E} = \boldsymbol{C}_e(\boldsymbol{R} - \boldsymbol{x}(k)) \tag{32}$$

where $\boldsymbol{C}_e = [\boldsymbol{c}, 1]$

$$\begin{aligned} \boldsymbol{s}(k+1) &= \boldsymbol{C}_e(\boldsymbol{R}_n - \boldsymbol{x}(k+1)) = \boldsymbol{C}_e(\boldsymbol{R}_1 - \boldsymbol{A}\boldsymbol{x}(k) - \boldsymbol{B}\boldsymbol{u}(k)) \\ &= \boldsymbol{C}_e\boldsymbol{R}_1 - \boldsymbol{C}_e\boldsymbol{A}\boldsymbol{x}(k) - \boldsymbol{C}_e\boldsymbol{B}\boldsymbol{u}(k) \end{aligned} \tag{33}$$

Control law can be written as follows:

$$\boldsymbol{u}(k)=(\boldsymbol{C}_e\boldsymbol{B})^{-1}(\boldsymbol{C}_e\boldsymbol{R}_1-\boldsymbol{C}_e\boldsymbol{A}\boldsymbol{x}(k)-\boldsymbol{s}(k+1)) \tag{34}$$

Submit Eq. (33) to Eq. (34), the control law can be rewritten as:

$$\boldsymbol{u}(k)=(\boldsymbol{C}_e\boldsymbol{B})^{-1}(\boldsymbol{C}_e\boldsymbol{R}_1-\boldsymbol{C}_e\boldsymbol{A}\boldsymbol{x}(k)-\boldsymbol{s}(k)-\mathrm{d}\boldsymbol{s}(k)) \tag{35}$$

where $\mathrm{d}\boldsymbol{s}(k)=-\boldsymbol{\varepsilon}\boldsymbol{T}\mathrm{sgn}(\boldsymbol{s}(k))-\boldsymbol{q}\boldsymbol{T}\boldsymbol{s}(k)$.

4.1.4.3 Design of Kalman filter

The algorithm of Kalman filter can be defined as follows:

$$\boldsymbol{M}_n(k)=\frac{\boldsymbol{P}(k)\boldsymbol{C}^{\mathrm{T}}}{\boldsymbol{C}\boldsymbol{P}(k)\boldsymbol{C}^{\mathrm{T}}+\boldsymbol{R}} \tag{36}$$

$$\boldsymbol{P}(k)=\boldsymbol{A}\boldsymbol{P}(k-1)\boldsymbol{A}^{\mathrm{T}}+\boldsymbol{B}\boldsymbol{Q}\boldsymbol{B}^{\mathrm{T}} \tag{37}$$

$$\boldsymbol{P}(k)=(\boldsymbol{I}_n-\boldsymbol{M}_n(k)\boldsymbol{C})\boldsymbol{P}(k) \tag{38}$$

$$\boldsymbol{x}(k)=\boldsymbol{A}\boldsymbol{x}(k-1)+\boldsymbol{M}_n(k)(\boldsymbol{y}_v(k)-\boldsymbol{C}\boldsymbol{A}\boldsymbol{x}(k-1)) \tag{39}$$

Based on Eq. (36) ~ Eq. (39), the output of fully compliant parallel manipulator after filter can be obtained as:

$$\boldsymbol{y}_e(k)=\boldsymbol{C}\boldsymbol{x}(k) \tag{40}$$

4.1.4.4 Stability analysis of sliding model controller

Variable rate reaching law can be written as follows:

$$\dot{\boldsymbol{s}}=-\boldsymbol{\varepsilon}\|\boldsymbol{x}\|_1\mathrm{sgn}(\boldsymbol{s}) \tag{41}$$

where $\|\boldsymbol{x}\|_1=\sum_{i=1}^{4}|\boldsymbol{x}_i|$ is system state norm.

Discrete form of Eq. (41) is:

$$\boldsymbol{s}(k+1)-\boldsymbol{s}(k)=-\boldsymbol{\varepsilon}\boldsymbol{T}\|\boldsymbol{x}(k)\|_1\mathrm{sgn}(\boldsymbol{s}(k)) \tag{42}$$

As:

$$\begin{aligned}(\boldsymbol{s}(k+1)-\boldsymbol{s}(k))\mathrm{sgn}(\boldsymbol{s}(k))&=(-\boldsymbol{\varepsilon}\boldsymbol{T}\|\boldsymbol{x}(k)\|_1\mathrm{sgn}(\boldsymbol{s}(k)))\mathrm{sgn}(\boldsymbol{s}(k))\\&=-\boldsymbol{\varepsilon}\boldsymbol{T}\|\boldsymbol{x}(k)\|_1|\boldsymbol{s}(k)|<0\end{aligned}$$

So when the sampling period $\boldsymbol{T}$ is too short, and then $2c-\varepsilon T>0$ and $2-\varepsilon T>0$, there has:

$$
\begin{aligned}
(s(k+1)-s(k))\operatorname{sgn}(s(k)) &= (2s(k)-\varepsilon T\|x(k)\|_1 \operatorname{sgn}(s(k)))\operatorname{sgn}(s(k)) \\
&= 2|s(k)| - \varepsilon T\|x(k)\|_1 \\
&\leqslant 2c|x_1| + 2|x_2| - \varepsilon T(|x_1| + |x_2|) \\
&= (2c-\varepsilon T)|x_1| + (2-\varepsilon T)|x_2| > 0 \qquad (43)
\end{aligned}
$$

According to existence and reaching conditions of discrete sliding model control theory, the design controller is stable.

4.1.5 Experimental Simulations

In this section, we shall experimentally evaluate the effectiveness of the proposed control technique in the vibration control of the fully compliant parallel manipulator. The schematic diagram of the proposed fully compliant parallel manipulator in our lab is shown as Fig. 4.

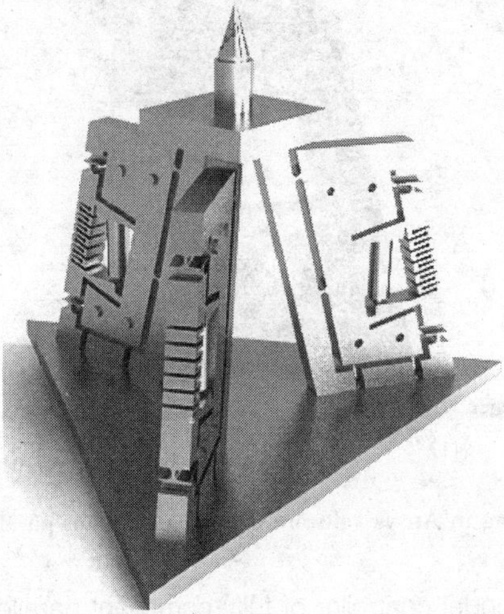

Fig. 4 Block diagram of the control system

4.1.5.1 Stiffness of fully compliant parallel manipulator

All elastic components are modeled as steel with elastic module, density and Poisson's ratio as $E = 210\text{GPa}$, $\rho = 7800\text{kg/m}^3$ and $\mu = 0.3$ respectively. Each flexure limb is divided into 9 beam elements and 35 generalized coordinate to meet the need of accuracy and boundary conditions. The actuators are made of PZT-5H piezoelectric ceramic, which its thickness is 0.5mm, length is 40mm and width is 20mm, piezoelectric constant, elastic module and density are 200×10^{-12} m/V, 1.2×10^{11} Pa and 7650kg/m^3 respectively. The allowable voltage of the piezoelectric actuators used in the system is from 0 to +150V. The sensors are resistance strain gage.

Meshing in Ansys software of fully compliant parallel manipulator is shown as Fig. 5, and natural frequency changes for the x and y axes of fully compliant parallel manipulator is shown in Fig. 6 and Fig. 7, respectively.

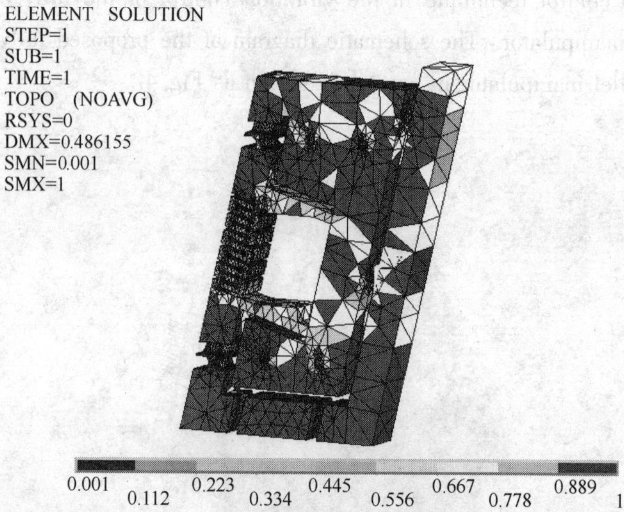

Fig. 5 Meshing in Ansys software of fully compliant parallel manipulator

4.1.5.2 Sliding model controller of fully compliant parallel manipulator

The gain of dynamic strain gauge is set to 1000, the brige voltage is 6V, cutoff frequency of filter is set to 100Hz, the sampling frequency of analog input and output is set to 1000Hz. The first two order mode is considered as the controlled one. The covariance of excitation signal is set to 100, and the covariance of measurement noise is estimated as 3×10^{-5}, the constant $\varepsilon = 0.01$ and the sampling period of the controller is 1ms, and the real-time

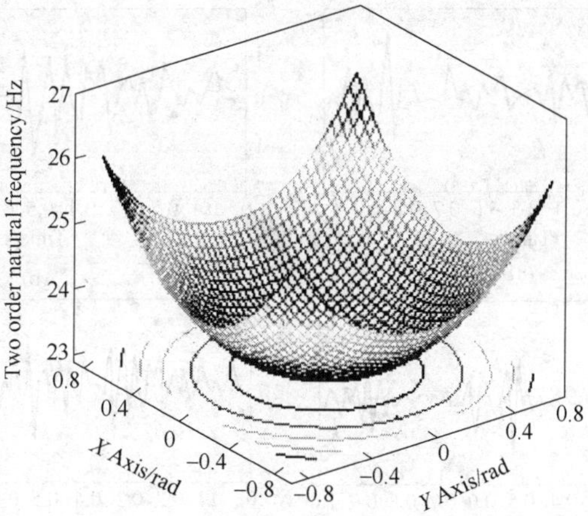

Fig. 6 Natural frequency changes for the x and y axes of fully compliant parallel manipulator

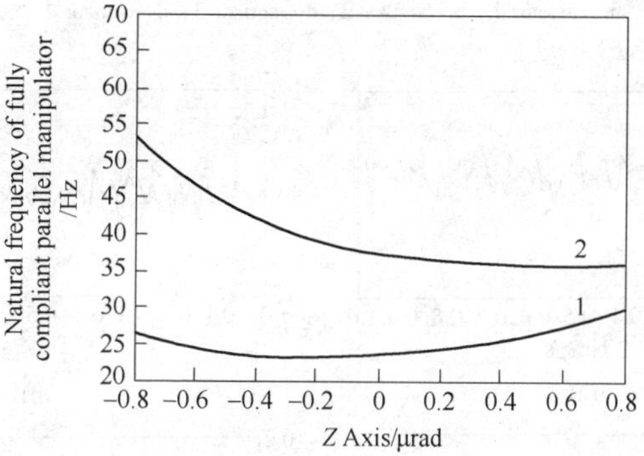

Fig. 7 Natural frequency changes for the z axis of fully compliant parallel manipulator
1—one order system; 2—two order system

control time is set to 1 s. Strain of each limb of 4-DoF compliant parallel manipulator is shown as Fig. 8 and Fig. 9, respectively.

It can be observed that the strains of sensors are remarkably reduced rapidly under the case where sliding model controller is adopted.

The partial node values for the displacement of the PZT actuators are shown in Table 1.

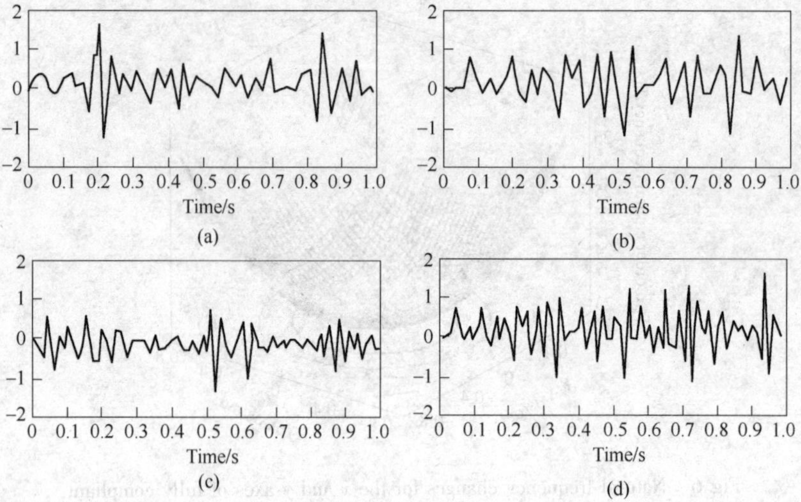

Fig. 8 Strain of the flexure limb from four sensor without controller

a—sensor 1; b—sensor 2; c—sensor 3; d—sensor 4

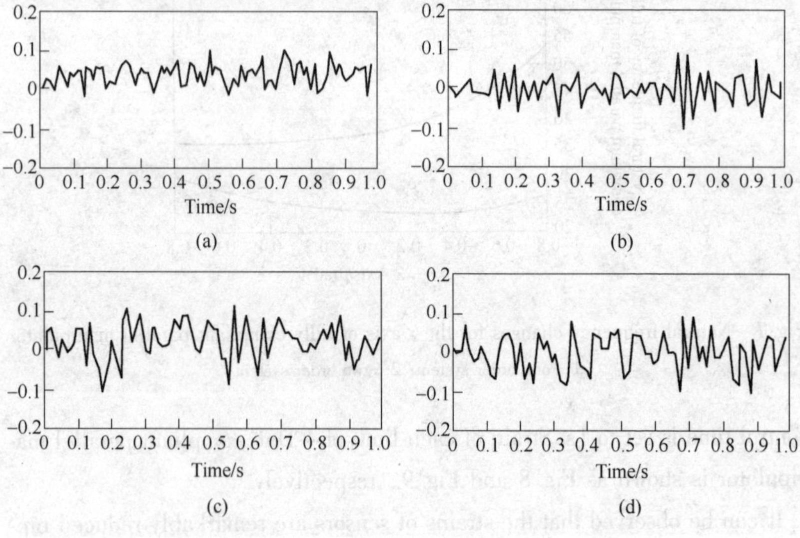

Fig. 9 Strain of the flexure limb from four sensor with sliding model controller

a—sensor 1; b—sensor 2; c—sensor 3; d—sensor 4

Table 1 Partial node values of displacement for the PZT actuators

NODE	*UX*	*UY*	*UZ*	*USUM*
33	0.368E-03	0.241E-04	0.746E-06	0.369E-03
96	0.351E-03	-0.152 E-04	0.706 E-06	0.352 E-03
438	0.315 E-03	-0.159 E-04	0.735 E-06	0.316 E-03
942	0.333 E-03	-0.166 E-04	0.653 E-06	0.333 E-03
2609	0.357 E-03	0.131 E-04	0.720 E-06	0.358 E-03
2805	0.331 E-03	-0.831 E-04	0.675 E-06	0.331 E-03
4098	0.298 E-03	-0.343 E-04	0.654 E-06	0.300 E-03
4152	0.307 E-03	-0.306 E-04	0.738 E-06	0.309 E-03
10759	0.306 E-03	0.478 E-05	0.664 E-06	0.306 E-03
14354	0.358 E-03	0.134 E-04	0.764 E-06	0.358 E-03

Impulse hammer method is applied to perform experimental modal test. The eZ-Analyst software is used to collect data of excitation point and all measurement points to obtain the frequency response function of the measuring point, and then export the data format which ME' scope VES can support. The ME' scope VES software is used to identify the modal parameters of the flexure mechanism. Frequency response fitting curve is shown in Fig. 10.

To verify the validation of the proposed sliding model controller, the covariance of excitation signal is set to 100, and the covariance of measurement noise is estimated as 3×10^{-6}.

By contrast to the results of Fig. 11 ~ Fig. 14, it can be observed that the deformations are remarkably reduced under the case where active vibration control with sliding model controller is effective to suppress the vibration response.

4.1.6 Conclusions

The vibration suppression of a 4-DoF fully compliant parallel manipulator equipped with piezoelectric actuator and strain gauge sensor has been inves-

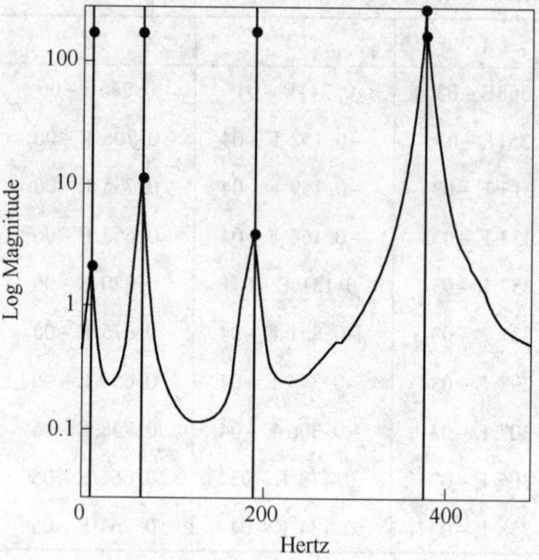

Fig. 10 Fitting curve of frequency response data of fully compliant parallel manipulator

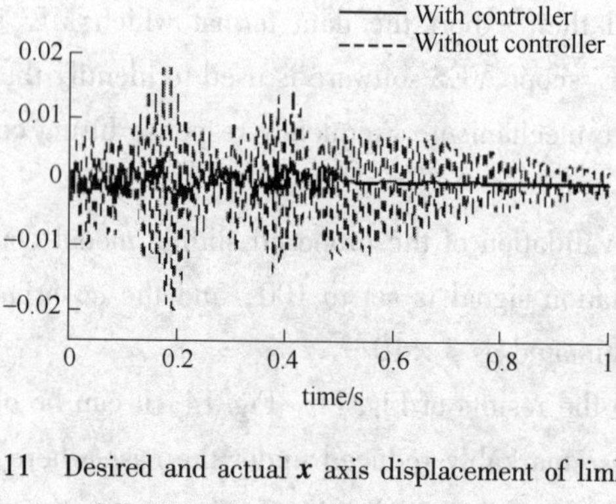

Fig. 11 Desired and actual ***x*** axis displacement of limb of fully compliant parallel manipulator

tigated by using discrete-time sliding mode control with disturbance observer. The differential kinematic/dynamic model is built, and then the stiffness of 4-DoF fully compliant parallel manipulator is analyzed. The discrete Kalman filtering technique is applied to construct the sliding mode control-

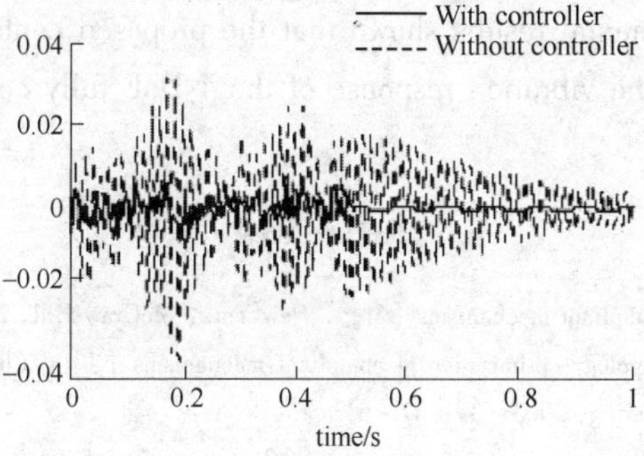

Fig. 12 Desired and actual **y** axis displacement of limb of fully compliant parallel manipulator

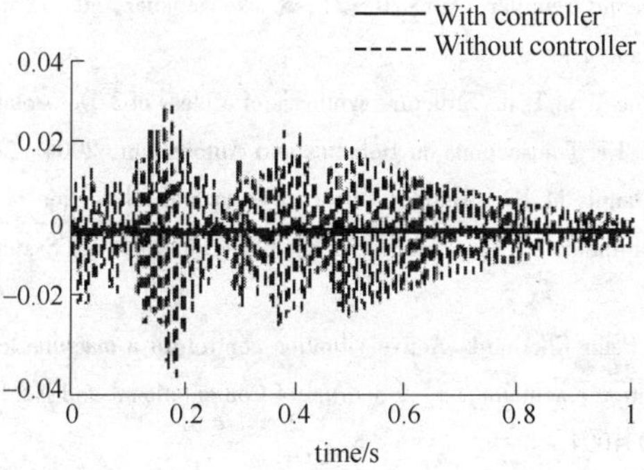

Fig. 13 Desired and actual *z* axis displacement of limb of fully compliant parallel manipulator

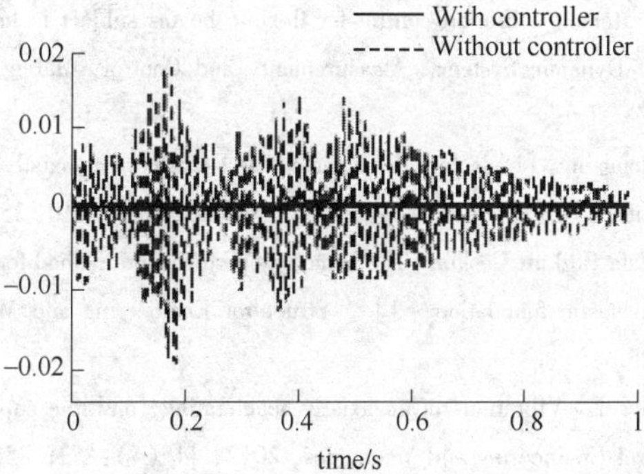

Fig. 14 Desired and actual rotational of limb of fully compliant parallel manipulator

ler. The experimental results shown that the proposed controller can effectively suppress the vibration response of the 4-DoF fully compliant parallel manipulator.

References

[1] Howell L L. Compliant mechanisms [M]. New York, McGraw-Hill, 2001.

[2] Zhang X M. Topology optimization of compliant mechanisms [J]. Chinese Journal of Mechanical Engineering, 2003, 39: 47 ~ 51.

[3] Janocha H. Adaptronics and smart structures: Basics, materials, design, and applications [M]. Springer, Berlin, 2007.

[4] Parallel kinematic machines in research and practice [C] //Proceedings of the Fourth Chemitz parallel kinematics Seminar, Apr. 20 ~ 21, R. Neugebauer, ed., Verlag Wissenschaftliche Scripten, Zwickau.

[5] Yuefa Fang, Lung-Wen Tsai. Structure synthesis of a class of 3-DoF rotational parallel manipulators [J]. IEEE Transactions on Robotics and Automation, 2004, 20 (1): 117 ~ 121.

[6] Boudewijn R, Dannis M. B., Meint J, et al. Design and fabrication of a planar three-DOFs MEMS-based manipulator [J]. Journal of Microelectromechanical Systems, 2010, 19 (5): 1116 ~ 1130.

[7] Alexandra Ast, Peter Eberhard. Active vibration control for a machine tool with parallel kinematics and adaptronic actuator [J]. Journal of Computational and Nonlinear Dynamics, July 2009, Vol. 4, 031004 - 1 - 8.

[8] Croft D, Shed G, Devasia S. Creep, Hysteresis, and Vibration compensation for Piezoactuators: Atomic Force Microscopy application [J]. Journal of Dynamic Systems, Measurement, and Control, 2001, 123: 35 ~ 43.

[9] Shang-The Wu. Remote vibration control for flexible beams subject to harmonic disturbances [J]. Journal of Dynamic Systems, Measurement, and Control, March 2004, 126: 198 ~ 238.

[10] Li Jun, Hua Hongxing, Li Xiaobin. Dynamic stiffness matrix of an axially loaded slender double-beam element [J]. Structural Engineering and Mechanics, 2010, 35 (6): 717 ~ 733.

[11] Baki Ozturk, Safa Bozkurt Coskun. The homotopy perturbation method for free vibration analysis of beam on elastic foundation [J]. Structural Engineering and Mechanics, 2011, 37 (4): 415 ~ 425.

[12] Kural S, Ozkaya E. Vibrations of an axially accelerating, multiple supported flexible beam [J]. Structural Engineering and Mechanics, 2012, 44 (4): 521 ~ 538.

[13] Qingsong Xu, Yangmin Li. Accuracy-based architecture optimization of a 3-DoF parallel kinematic machine [C] //Proceedings of the 2006 IEEE International Conference on Automa-

tion Science and Engineering, Shanghai, China, October 7 ~ 10, 2006: 63 ~ 68.

[14] Wu Huapeng, Heikki Handroos. Utilization of differential evolution in inverse kinematics solution of a parallel redundant manipulator [C] //Fourth International Conference on Knowledge-Based Intelligent Engineering System & Allied Technology, 30 August-1 September, 2000, Brighton, UK, 821 ~ 815.

[15] You-An Zhang, Yu-lin Mi, Ming Zhu, Feng-lin Lu. Adaptive sliding mode control for two-link flexible manipulators with H finite tracking performance [C] //Proceedings of the Fourth International Conference on Machine Learning and Cybernetics , Guangzhou, 18 ~ 21 August, 2005: 702 ~ 707.

[16] Dachang Zhu, Guoxin Zhang, Yuefa Fang. Neural-adaptive sliding mode control of 4-SPS (PS) type parallel manipulator [C] //2008 10th International Conference on Control, Automation, Robotics and Vision, 17 ~ 20 December, 2008, Hanoi, Vietnam, 2008: 2055 ~ 2059.

[17] Shiferaw Dereje, Jain, Anamika. Comparision of joint space and task space integral sliding mode controller implementations for 6-DoF parallel robot [C] //11th WSEAS International Conference on Robotics, Control and Manufacturing Technology, 8 ~ 10 March, 2011, Venice, Italy, 163 ~ 169.

4.2 Vibration Control of Smart Structure Using Sliding Mode Control with Observer

This paper studies the application of the sliding mode control method to reduce the vibration of flexible structure with piezoelectric actuators and strain gage transducer in practical complex environment. The state-space dynamic model of the system was derived by using finite element method and experimental modal test. The structure is subjected to arbitrary, unmeasurable disturbance forces. Taking into account the uncertain random disturbance and measurement noise, Kalman filter is chosen as the state estimator to obtain the modal coordinates and modal velocities for the modal space control. A sliding mode controller is adopted due to its distinguished robustness property of insensitiveness to parameter uncertainties and external disturbances. The sliding surface is determined by using optimization method, and the sliding controller is designed by applying Lyapunov direct method. That is, along the switching surface, the cost function of the states is minimized. A real-time control system was built using dSPACE DS1103 platform, and vibration control tests were performed to experimentally verify the performances of the proposed controller. The results of experiment show the controller can effectively attenuate elastic vibration of the structure.

4.2.1 Introduction

In precision and aerospace industry, many researches on lightweight and miniaturized structures have been carried out to improve structural performances. Among the researches, passive structures using composite material are typically known as one of the effective methods. However, the traditional passive structures are very sensitive to change of internal load condition and external environment condition which can even cause sudden destruction of structures. Therefore, in order to satisfy stringent requirements for precision control and lightweight miniaturization, smart materials such as shape memory alloys, piezoceramics, electroorheological fluids and magnetorheological fluids are frequently adopted for smart structures. The performance requirements of future space structures, jet fighters and concept automobiles have brought much interest to the area of smart structures. A smart structure can be defined as a structure with bonded or embedded sensors and actuators as well as an associated control system, which enable the structure to respond simultaneously to external stimuli exerted on it and then suppresses undesired effects or enhance desired effects. Among various smart structures, those with piezoelectric patches have received much attention in recent years, due to the fact that piezoelectric materials have simple mechanical properties, small volume, lightweight, large useful bandwidth, efficient conversion between electrical and mechanical energy, good ability to perform vibration control and ease of integration into metallic and composite structures[1-3]. Different control techniques have been investigated in the control of smart structure. Abreu conducted experimental work for the vibration control of flexible beam by using piezoelectric sensors and actuators with Linear Quadratic Gaussian (LQG) controller[4]. There are many classical strategies that can be used when the mathematical model is available, for instance pole allocation and optimal control. However, if the model has uncertainties these methods are not indicated. There are many robust techniques in structural control literature. Li investigated two control strategies for robust vibration control of parameter uncertain systems[5]. Mayhan combined intelligent control and smart materials to produce an adaptive and robust controller to dampen the fundamental vibration mode of the system in the presence of modeling uncertainties[6]. Zhang et al. studied the active vibration control problem for the high-speed flexible mechanisms all of whose members were considered as flexible by using complex mode method and robust H_∞ control scheme[7~8]. Kawabe utilized neural networks (NN) theory for active control in a longitudinal cantilevered-beam system by simulation and experiment. It is found that fairly satisfactory active damping effect using the NN controller is obtained[9]. But the random disturbance and measurement noise of the actual system were not considered by these currently proposed vibration control strategies. The issue of robustness against external disturbances was not addressed, and therefore the proposed vibration controllers cannot be effectively applied to the smart structure under the random uncertain disturbances. Because sliding mode control has inherent robustness to system parameter variation and external disturbances, it is meaningful to investigate its application in vibration control of smart structure. We aim here to deal with the active vibration reduction problem in flexible structure with uncertainties through designing reasonable sliding mode controller. The developed control strategy integrates the sliding mode control strategy and Kalman filter technique. In this paper, the vibration control of a flexible beam is investigated by using sliding mode control with observer and experimental modal test method, and taking into account the random disturbance uncertainty, modal parameter uncertainty and measurement noise. The paper is organized as follows. In section 2, a dynamic model of a flexible beam with piezoelectric actuators and strain gauge sensors is constructed by using finite element method. In section 3, the sliding mode controller with observer is proposed, and sliding surface is determined by using optimization method, and the sliding controller is designed by applying Lyapunov direct method. In section 4, experimental identification of the flexible cantilever beam is performed to obtain its modal parameters. And the experimental validation test is performed based on the dSPACE DS1103 platform. The conclusions are given in section 6.

4.2.2 Dynamic Modeling of Smart Structure

The modeling of smart structure with piezoelectric actuators and sensors has been a subject of intense research for a long time and is only briefly described here.

The flexible structure is modeled by using a two-node beam element. The beam element is shown in Fig. 1, which has two nodes with four degrees of freedom at each node; namely u_1, u_5, the longitudinal displacement, and u_2, u_6, the transverse displacement, and u_3, u_7, the slope, and u_4, u_8, the curvature. L_e is the length of element. The nodal displacement vector u with respect to reference frame $A-\overline{xy}$ is expressed as

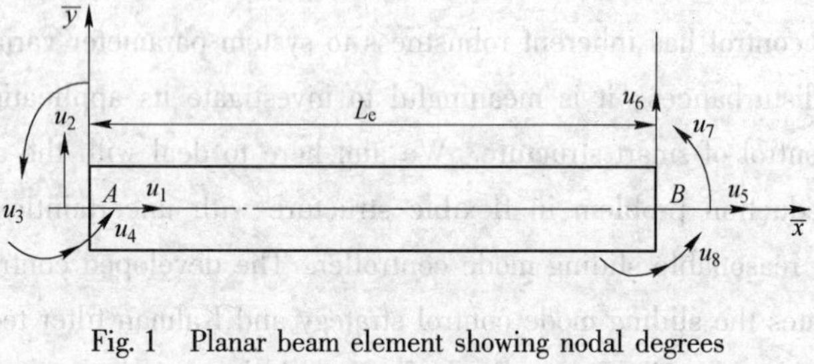

Fig. 1 Planar beam element showing nodal degrees of freedom and coordinate systems

$$\boldsymbol{u} = [u_1 \quad u_2 \quad u_3 \quad u_4 \quad u_5 \quad u_6 \quad u_7 \quad u_8]^{\mathrm{T}}$$

The transverse and longitudinal displacement fields of two-node beam element are constructed using the quintic hermite and linear interpolation polynomials, respectively. V, W denotes longitudinal and transverse elastic displacement of arbitrary point, respectively. Subscript T denotes matrix transpose. They can be written by the following form

$$\begin{bmatrix} V(\bar{x},t) \\ W(\bar{x},t) \end{bmatrix} = \begin{bmatrix} N_1(\bar{x}) \\ N_2(\bar{x}) \end{bmatrix} \boldsymbol{u}$$

where $\boldsymbol{N}(\bar{x}) = [N_1(\bar{x}) \quad N_2(\bar{x})]^{\mathrm{T}}$ is shape function, $\boldsymbol{u}$ is the nodal displacement vector.

The system dynamic equations can be obtained by using finite element method

$$\boldsymbol{M}\ddot{\boldsymbol{U}} + \boldsymbol{C}\dot{\boldsymbol{U}} + \boldsymbol{K}\boldsymbol{U} = \boldsymbol{P} \tag{1}$$

where $\boldsymbol{M}$, $\boldsymbol{C}$, $\boldsymbol{K}$ are the systematic mass, damping, stiffness, respectively. $\boldsymbol{U}$, $\dot{\boldsymbol{U}}$, $\ddot{\boldsymbol{U}}$ are the generalized displacement, velocity, and acceleration vectors of the system, respectively. $\boldsymbol{P}$ is the systematic generalized disturbance force vector corresponding to vector $\boldsymbol{U}$.

The piezoelectric patches work as actuators are perfectly bonded on the upper and lower surfaces of the beam at the same location. For the modeling of PZT actuators, literature[10, 11] provides a detailed derivation of coupling of PZT actuators and a host beam. These bending moments induced by the actuators is given by

$$T_A = T_B = -d_{31}E_p b_p (t_p + t_a)\boldsymbol{V}_{\text{in}} \tag{2}$$

where d_{31}, E_p, b_p, t_p is piezoelectric constant, elastic module, width, and thickness of PZT patch, respectively. t_a is the thickness of the beam element, $\boldsymbol{V}_{\text{in}}$ is the vector of input voltage to the piezoelectric actuators. The moments are assembled as a part of the external moments exerted on node.

Strain is the amount of deformation of a structure due to an applied force. Strain gauge is the most common sensor for measuring strain. A strain gauge's electrical resistance varies in proportion to the amount of strain placed on it. The deformations mainly include compression or tension and bending deformation for the flexible beam. The strain ε in $\bar{x}$ direction is given by

$$\varepsilon = \varepsilon_L + \varepsilon_B \tag{3}$$

where ε_L is compression or tension strain and ε_B is longitudinal strain due to bending deformation, respectively. They can be given by

$$\varepsilon_L = N_1''u = N_1''\boldsymbol{B}U,\ \varepsilon_B = -hN_2''u = -hN_2''\boldsymbol{B}U \tag{4}$$

where h is the distance between the neutral axis of the beam and the outer surface of the beam, N_1'', N_2'' are the first-order and the second-order differential of the shape function N_1 and N_2, respectively, $\boldsymbol{B}$ denotes the transformation matrix. Thus, the exogenous perturbation and the control inputs have no direct effect on the measured outputs.

There are a piezoelectric actuators and s strain gage sensors on the flexible structure. Combining Eq. (1) ~ Eq. (4), the dynamic equations of the structure equipped with piezoelectric actuators and strain gauge transducer can be expressed as

$$\boldsymbol{M}\ddot{\boldsymbol{U}} + \boldsymbol{C}\dot{\boldsymbol{U}} + \boldsymbol{KU} = \boldsymbol{P} + \boldsymbol{D}_a\boldsymbol{V}_{\text{in}},\ y = \boldsymbol{D}_s\boldsymbol{U} \tag{5}$$

where $\boldsymbol{D}_a$ is the systematic control matrix related to configuration of actuators, $\boldsymbol{D}_s$ is the systematic output matrices determined by configuration of sensors, a a-by-1 vector, and y is the strain from the sensors, a s -by-1vector.

It was shown that the dynamic response of the flexible structure is composed mainly of the lower modes. In order to control the lower modes, the physical-coordinate equations must be first transformed into modal coordinates. Here, we choose the first c order modes as control modes. Applying the modal theory, the normalized modal transformation is introduced by

$$\boldsymbol{U} = \boldsymbol{\psi}_e\boldsymbol{\eta}_e \tag{6}$$

where $\boldsymbol{\psi}_e$ is the controlled normalized modal matrix, $\boldsymbol{\eta}_e$ is controlled modal coordinate vectors. Substituting Eq. (6) into Eq. (5), the system dynamic equations are rewritten as

$$\ddot{\boldsymbol{\eta}}_e + \boldsymbol{C}_e\dot{\boldsymbol{\eta}}_e + \boldsymbol{K}_e\boldsymbol{\eta}_e = \boldsymbol{N}_e + \boldsymbol{D}_{ac}\boldsymbol{V}_{\text{in}},\ \boldsymbol{y} = \boldsymbol{D}_{se}\boldsymbol{\eta}_e \tag{7}$$

where $\boldsymbol{N}_e = \boldsymbol{\psi}_e^{\mathrm{T}}\boldsymbol{P}$, $\boldsymbol{D}_{ac} = \boldsymbol{\psi}_e^{\mathrm{T}}\boldsymbol{D}_a$, $\boldsymbol{D}_{se} = \boldsymbol{\psi}_e^{\mathrm{T}}\boldsymbol{D}_a$, $\boldsymbol{C}_e$, $\boldsymbol{K}_e$ are $e \times e$ diagram matrix, which is determined by system natural frequency and damping ratio.

For control synthesis, the system must be written as a system of first-order ordinary differential equations (ODEs) . We can define the controlled state variables by the following form

$$\boldsymbol{X}_e = [\eta_1 \quad \cdots \quad \boldsymbol{\eta}_e \quad \dot{\eta}_1 \quad \cdots \quad \dot{\boldsymbol{\eta}}_e]^{\mathrm{T}}$$

Due to the controlled mode number e, the number of controlled state variables is $2e$. Taking into account measurement noise v, the state-space model for the system can be written as

$$\dot{X}_e = A_e X_e + B_e V_{in} + N_e,\ y = C_e X_e + v \tag{8}$$

where $A_e = \begin{bmatrix} \mathbf{0}_e & I_e \\ -K_e & -C_e \end{bmatrix}$, $B_e = \begin{bmatrix} \mathbf{0}_{ae} \\ D_{ae} \end{bmatrix}$, $C_e = [D_{se} \quad \mathbf{0}_{se}]$, I_e is $e \times e$ unit matrix, $\mathbf{0}_{ae}$, $\mathbf{0}_{se}$ are $e \times a$, $s \times e$ zero matrix, respectively.

4.2.3 Control System Design

Sliding mode controller (SMC) is a function of more than two structures and gives some desirable closed-loop properties. The desirable features include invariance, order reduction, and robustness against parameter variations and disturbances. The stability and robustness of SMC is guaranteed using the concept of switching function and Lyapunov stability theory. Sliding mode control is to design a controller such that the motion of the system tends to sliding mode surface. Therefore, the design of SMC includes the determination of sliding surface and controller design. The block diagram of the control system is shown in Fig. 2. In the figure, $\hat{X}_e$ is the estimated value of state vector X_e by using Kalman filter. And the measurement noise v is considered during the controller design.

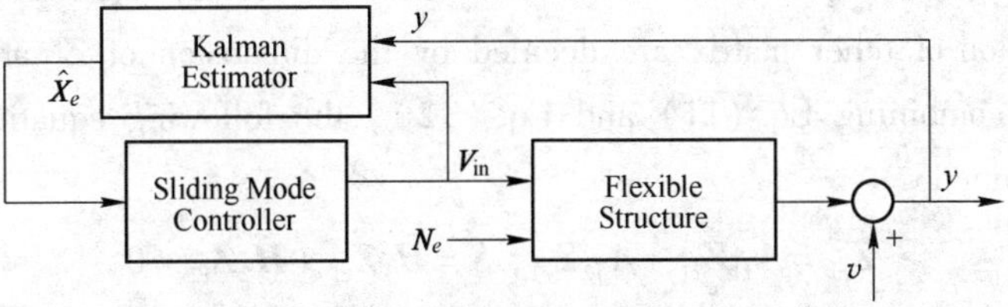

Fig. 2 Block diagram of the control system

4.2.3.1 Sliding mode surface design

The problem now is to determine a sliding surface which guarantees stable sliding mode motion on the surface itself.

The sliding surface S can be expressed as

$$\boldsymbol{S} = \boldsymbol{H}\boldsymbol{X}_e = 0 \tag{9}$$

where $\boldsymbol{S} = [S_l \quad S_l \quad \cdots \quad S_a \quad]^{\mathrm{T}}$, S_l is the sliding variable, $\boldsymbol{H}$ is $a \times 2c$ matrix. In order to determine the matrix $\boldsymbol{H}$, the state of the system is to make the following linear transformation.

$$\boldsymbol{Z}_c = \boldsymbol{\Gamma}\boldsymbol{X}_c \tag{10}$$

where $\boldsymbol{\Gamma}$ is the transformation matrix, it can be given by

$$\boldsymbol{\Gamma} = \begin{bmatrix} \boldsymbol{I}_1 & -\boldsymbol{B}_1\boldsymbol{B}_2^{-1} \\ 0 & \boldsymbol{I}_2 \end{bmatrix}, \ \boldsymbol{B}_c = \begin{bmatrix} \boldsymbol{B}_1 \\ \boldsymbol{B}_2 \end{bmatrix}$$

where $\boldsymbol{I}_1$ and $\boldsymbol{I}_2$ are $(2c-a) \times a$ and $a \times a$ unit matrix, $\boldsymbol{B}_1$ and $\boldsymbol{B}_2$ are $(2c-a) \times a$ and $a \times a$ matrix, respectively, and $\boldsymbol{B}_2$ is nonsingular matrix.

Substituting Eq. (10) into Eq. (8), the state equation and sliding surface can be rewritten as the following form expressed by $\boldsymbol{Z}_c$.

$$\dot{\boldsymbol{Z}}_c = \boldsymbol{\Gamma}\boldsymbol{A}_c\boldsymbol{\Gamma}^{-1}\boldsymbol{Z}_c + \boldsymbol{\Gamma}\boldsymbol{B}_c\boldsymbol{V}_{\mathrm{in}} = \tilde{\boldsymbol{A}}\boldsymbol{Z}_c + \tilde{\boldsymbol{B}}\boldsymbol{V}_{\mathrm{in}}, \ \boldsymbol{S} = \tilde{\boldsymbol{H}}\boldsymbol{Z}_c = 0 \tag{11}$$

where $\tilde{\boldsymbol{A}} = \boldsymbol{\Gamma}\boldsymbol{A}_c\boldsymbol{\Gamma}^{-1}$, $\tilde{\boldsymbol{B}} = [0 \quad \boldsymbol{B}_2^{\mathrm{T}}]^{\mathrm{T}}$, $\tilde{\boldsymbol{H}} = \boldsymbol{H}\boldsymbol{\Gamma}^{-1}$. The Eq. (11) can be partitioned to yield

$$\boldsymbol{Z}_c = \begin{bmatrix} \boldsymbol{Z}_{c1} \\ \boldsymbol{Z}_{c2} \end{bmatrix}, \ \tilde{\boldsymbol{A}} = \begin{bmatrix} \tilde{\boldsymbol{A}}_{11} & \tilde{\boldsymbol{A}}_{12} \\ \tilde{\boldsymbol{A}}_{21} & \tilde{\boldsymbol{A}}_{22} \end{bmatrix}, \ \tilde{\boldsymbol{H}} = [\tilde{\boldsymbol{H}}_1 \quad \tilde{\boldsymbol{H}}_2] \tag{12}$$

where $\boldsymbol{Z}_{c1}$ and $\boldsymbol{Z}_{c2}$ are $2c-a$ and a dimensional vector, respectively, the dimension of other matrix are decided by the dimension of $\boldsymbol{Z}_{c1}$ and $\boldsymbol{Z}_{c2}$. Thus, combining Eq. (11) and Eq. (12), the following equations can be obtained

$$\dot{\boldsymbol{Z}}_{c1} = \tilde{\boldsymbol{A}}_{11}\boldsymbol{Z}_{c1} + \tilde{\boldsymbol{A}}_{12}\boldsymbol{Z}_{c2}, \ \boldsymbol{S} = \tilde{\boldsymbol{H}}_1\boldsymbol{Z}_{c1} + \tilde{\boldsymbol{H}}_2\boldsymbol{Z}_{c2} = 0 \tag{13}$$

In order to simplify the design of sliding surface, we can assume

$$\tilde{\boldsymbol{H}}_2 = \boldsymbol{I} \tag{14}$$

where $\boldsymbol{I}$ is unit matrix. Therefore, combining Eq. (13) and Eq. (14), the following equations can be obtained where $\boldsymbol{Z}_{c1}$ is considered as state variables, $\boldsymbol{Z}_{c2}$ is output variables.

$$\dot{Z}_{c1} = (\tilde{A}_{11} - \tilde{A}_{12}\tilde{H}_1)Z_{c1},\ Z_{c2} = -\tilde{H}_1 Z_{c1} \tag{15}$$

Using the optimal method, the matrix $\boldsymbol{H}$ is determined by minimizing the following performance index J.

$$J = \int_{t0}^{\infty} X_c^{\mathrm{T}} R X_c \mathrm{d}t \tag{16}$$

where $\boldsymbol{R}$ is positive define weighting matrix. Considering the physical meaning of state variable X_c, the weighting matrix can be selected by the following expression.

$$R = \begin{bmatrix} K_c & -M_c \\ -M_c & M_c \end{bmatrix}$$

where M_c, K_c are the mass matrix and stiffness matrix corresponding to the first c control modes, respectively. Substituting Eq. (10) into Eq. (16), the performance index function can be written by the form of state variable Z_c.

$$J = \int_{t0}^{\infty} [Z_{c1}^{\mathrm{T}} \quad Z_{c2}^{\mathrm{T}}] \mathrm{T} \begin{bmatrix} Z_{c1} \\ Z_{c2} \end{bmatrix} \mathrm{d}t \tag{17}$$

where

$$T = (\Gamma^{-1})^{\mathrm{T}} R \Gamma^{-1} = \begin{bmatrix} T_{11} & T_{12} \\ T_{21} & T_{22} \end{bmatrix}$$

Combining Eq. (15) and Eq. (17), the solution of the performance function is a linear quadratic optimization problem. Using the Linear Quadratic (LQ) algorithm, the optimization problem can be solved and the following expression can be obtained

$$Z_{c2} = -K_0 Z_{c1} \tag{18}$$

Comparing Eq. (18) and Eq. (15), we can obtain

$$\tilde{H}_1 = K,\ H = [K_0 \quad I]\Gamma \tag{19}$$

Substituting Eq. (19) into Eq. (9), the sliding surface can be obtained.

4.2.3.2 Sliding mode controller

Lyapunov direct method is applied to design the sliding mode controller such that the response of the system can tend to the sliding surface expressed in Eq. (9) by determining the control input. Suppose the Lyapunov function is defined as

$$V = \frac{1}{2}\boldsymbol{S}^{\mathrm{T}}\boldsymbol{S} = \frac{1}{2}\boldsymbol{X}_c^{\mathrm{T}}\boldsymbol{H}^{\mathrm{T}}\boldsymbol{H}\boldsymbol{X}_c \tag{20}$$

According to Lyapunov asymptotic stability condition and Eq. (20), we can obtain

$$\dot{V} = \boldsymbol{S}^{\mathrm{T}}\dot{\boldsymbol{S}} \leqslant 0 \tag{21}$$

where $\dot{\boldsymbol{S}} = \boldsymbol{H}\dot{\boldsymbol{X}}_c$. Substituting Eq. (8) into Eq. (21), the above equation can be rewritten as

$$\dot{V} = \boldsymbol{S}^{\mathrm{T}}\boldsymbol{H}\dot{\boldsymbol{X}}_c = \boldsymbol{S}^{\mathrm{T}}\boldsymbol{H}(\boldsymbol{A}_c\boldsymbol{X}_c + \boldsymbol{B}_c\boldsymbol{V}_{in} + \boldsymbol{N}_c) = \boldsymbol{\chi}(\boldsymbol{V}_{\mathrm{in}} - \boldsymbol{G})$$

where $\boldsymbol{\chi} = \boldsymbol{S}^{\mathrm{T}}\boldsymbol{H}\boldsymbol{B}_c$, $\boldsymbol{G} = -(\boldsymbol{H}\boldsymbol{B}_c)^{-1}(\boldsymbol{H}\boldsymbol{A}_c\boldsymbol{X}_c + \boldsymbol{H}\boldsymbol{N}_c)$.

In order to satisfy the Lyapunov stable condition $\dot{V} \leqslant 0$, the control voltages can be selected as

$$\boldsymbol{V}_{\mathrm{in}} = \boldsymbol{G} - \boldsymbol{\delta}\boldsymbol{\chi}^{\mathrm{T}} \tag{22}$$

where $\boldsymbol{\delta}$ is diagonal matrix whose diagonal elements $\boldsymbol{\delta} \geqslant 0$. Thus, the expression $\dot{V} = \boldsymbol{\chi}\boldsymbol{\delta}\boldsymbol{\chi}^{\mathrm{T}} \leqslant 0$ can be satisfied. Meanwhile, taking account into the bound of control voltage, the control inputs are set as.

$$V_i = \begin{cases} V_i^* & |V_i^*| \leqslant V_m \\ V_m \operatorname{sgn}(V_i^*) & |V_i^*| > V_m \end{cases} \tag{23}$$

where subscript i denotes the ith control input voltage, V_i^* is the ith voltage computed by the controller, V_i is the actual voltage applied to the actuator, V_m is the bound of the control voltage, sgn(·) represents sign function.

4.2.3.3 Kalman state estimator

Since the modal positions and velocities of the system are not directly measurable, a state observer may be employed to obtain the estimated states. The Kalman filter can provide an efficient way to estimate the state of a system, in a way that minimizes the mean of the squared error. Considering external disturbance and measurement noise, we can design a state estimator given the state-space model of the system shown in Eq. (8) by using Kalman technology.

Here, suppose the exogenous disturbance and measurement noise satisfy the following conditions

$$E(N_c)=E(v)=0,\ E(N_cN_c^{\mathrm{T}})=Q_d,\ E(vv^{\mathrm{T}})=R_n \tag{24}$$

where $E(\cdot)$ denotes the expected value of a variable, Q_d and R_n are co-variance of the external disturbance and measurement noise, respectively. We can construct a state estimate $\tilde{X}_c$ that minimizes the steady-state error covariance

$$P=\lim_{t\to\infty}E(\{X_c-\hat{X}_c\}\{X_c-\hat{X}_c\}^{\mathrm{T}})$$

The optimal solution is the Kalman filter with the following state equations.

$$\dot{\hat{X}}_c=A_c\hat{X}_c+B_cV_{\mathrm{in}}+L(y-C_c\hat{X}_c),\ \hat{y}=C_c\hat{X}_c$$

where the filter gain matrix L is determined by solving corresponding algebraic Riccati equation, $\hat{y}$ is estimated value of y. The estimator uses the known inputs V_{in} and the measurements y to generate the output and state estimates as shown in Fig. 2.

4.2.4 Experimental Investigation

In this section, we shall experimentally evaluate the effectiveness of the proposed control method in the vibration control of a flexible beam.

4.2.4.1 Experimental setup

The length of the cantilever beam is 300mm, its width and thickness are 20mm, 1.5mm, respectively. Its material is steel with elastic module, density and Poisson's ratio 2100GPa, 7800kg/m^3, 0.3, respectively. The beam is divided into 7 beam elements and 28 generalized coordinates shown as in the Fig. 3. The lengths of the elements are 50 mm, 40 mm, 40 mm, 40 mm, 40 mm, 40 mm, 50 mm, respectively. The symbols N, E, A, S denote node, actuator, sensor, respectively, in the Fig. 3. The actuators are made of PZT - 5H piezoelectric ceramic, which its thickness is 0.5 mm, length is 40 mm and width is 20 mm, piezoelectric constant d_{31}, elastic module and density is 200×10^{-12} m/V, 2100GPa, 7650kg/m^3, respectively. The sensors are resistance strain gage. The configuration of the actuators and sensors are shown in Fig. 3. The three pairs actuator bonded on the link are located at elements E_2, E_4 and E_6. Two sensors S_1, S_2 are located at the midpoint of elements E_3 and E_5. Three actuators A_1, A_2, A_3 are located at elements E_2, E_4 and E_6, respectively.

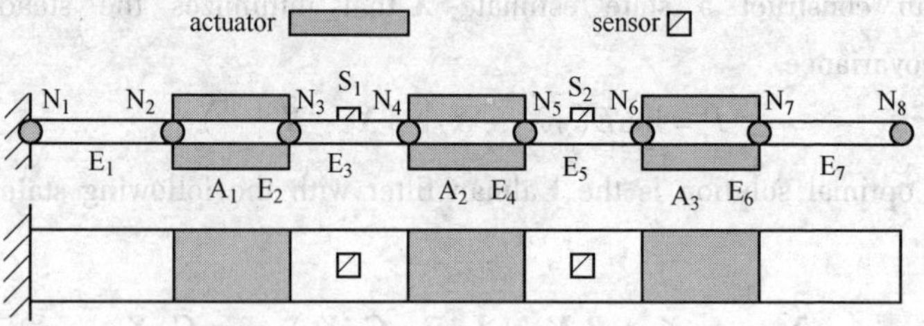

Fig. 3 Configuration of the elements and nodes, actuator and sensor, N, E, A, S denote node, actuator, sensor, respectively

Fig. 4 illustrates the experimental setup that consists of a flexible beam bonded on PZT and strain gauge. By the way, the experimental device is also used to study the vibration control of high-speed 5-bar mechanism. As the bolt is tighten, the rotational pair becomes a fixed pair, that is, the flexible link can be considered as a cantilever beam. The high-speed analog input and output ports are provided with dSPACE DS1103. The electric amplifier made in Harbin core tomorrow science and technology Co., Ltd is used to drive piezoelectric patch. Resistance strain gauges are made in ZEMIC Co., Ltd, its type is BE120-3AA (11), the resistance is 120Ω, and the sensitivity coefficient is 2.17. The strain gauge is connected to dynamic strain gauge through 1/4 electric bridge which is used to transform the strain signal to voltage.

4.2.4.2 Experimental model test

Experimental modal analysis is a case of system identification where a priori model form consisting of modal parameters is assumed. Because it is hard to obtain damping ratio of structure by finite element method (FEM), the

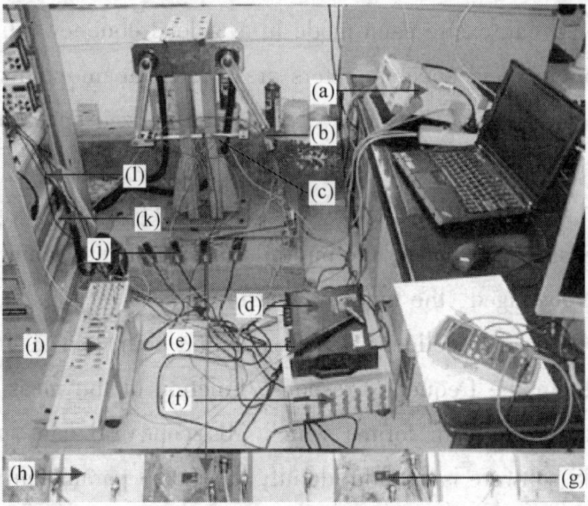

Fig. 4 Experimental setup

(a) signal generator; (b) cantilever beam; (c) accelerometer; (d) dynamic signal acquisition; (e) hammer; (f) dynamic strain gauge; (g) strain gauge; (h) piezoelectric patch; (i) DS1103 connector panel; (j) electric bridge box; (k) voltage amplifier; (l) industrial computer

experimental modal test is a good method to get accurate natural frequency and damping ratio, which provide a basis for adjustment of the control model of the flexible structure. The setup of the experimental modal test is shown as Fig. 4. Kistler 8690C50-type piezoelectric accelerometer is used as acceleration sensor. The ZonicBook/618E is the dynamic signal acquisition system. The eZ-Analyst software is real-time vibration analysis software equipped with a dynamic signal acquisition system which provides a real-time analysis capability in the frequency domain and time domain. ME'scopeVES software is used to be a post-processing test which is capable of analyzing mechanical and structural static and dynamic characteristics to obtain modal parameters.

Impulse hammer method is applied to perform experimental modal test. To excite the bending vibration, the cantilever beam was hit with a hammer at the specified excitation points. Using a hammer to produce a wide band of excitation, it can excite each mode in a wider frequency range. The locations of hammer and accelerometers are located at midpoints of element E_3 and elements $E_1 \sim E_7$, respectively. A miniaturized accelerometer PCB is sequentially placed at different locations. The tap position of the hammer is fixed, and the measurement points are 7 different positions. In order to eliminate measurement noise, the multiple measurements value of measurement point is averaged, the times of measurement is set to 5. The eZ-Analyst software is used to collect data of excitation point and all measurement points, to obtain the frequency response function of the measuring point, and then export the data format which ME'scopeVES can support. The ME'scopeVES software is used to identify the modal parameters of the flexible beam. The synthesis of the controller is based on a nominal model constructed by low-frequency modes. In the present case, the first four-order mode of the link is identified. The first four-order nature frequencies by using experimental identification and FEM are shown in Table 1, respectively, and the corresponding damping ratios are identified and tabulated in Ta-

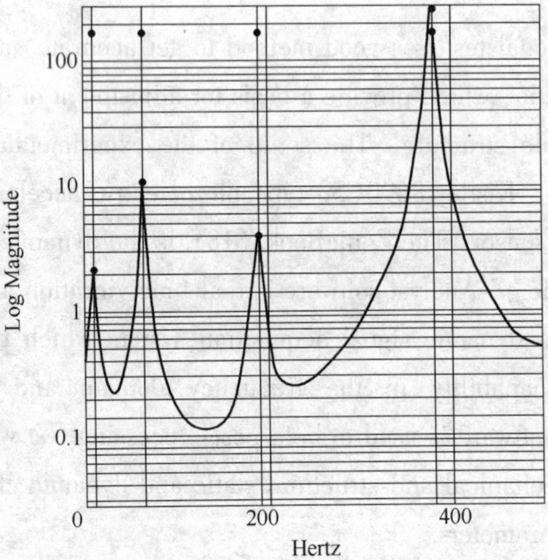

Fig. 5 Fitting curve of frequency response data

ble 1. As can be seen from the table, relative error of calculated and experimental values of natural frequency is close to 2% ~ 6%, which indicates that the finite element model is not fully consistent in the actual system. That is to say, the model used to design controller is uncertain. Fitting curve of frequency response data is shown as Fig. 5.

Table 1 The First Four-order Natural Frequency and Damping Ratio

Mode order	Natural frequency/Hz		Relative error	Damping ratio (%)
	Calculation value	Experimental value		
1	10.5	10.8	1.52	1.52
2	73.8	72.6	2.37	2.37
3	180.9	189.6	4.56	1.24
4	366.7	387.5	5.37	0.96

4.2.4.3 Experimental results

Schematic diagram of vibration control experiment is shown as Fig. 6. The principal of the vibration control is described as follows. When the exogenous disturbances are exerted on the beam, the vibration response will be generated. The output of the strain gage is given as input to the dynamic strain gauge which filters out the noise contents. The conditioned sensor signal is given as analog input card through the electric bridge connector box. The vibration signal measured by a sensor is transformed into a voltage signal by a dynamic strain transducer, and through a low-pass filter and an A/D converter, the analog voltage signal is converted into a digital signal to the dSPACE controller board. And the control voltage applied to actuator can be obtained through the designed controller. As the control voltage from D/A port is relatively low, the voltage exerted on the piezoelectric patch must be amplified by the voltage amplifier to implement the vibration control. The control signal calculated by the dSPACE is converted into an analog signal by a D/A converter, and then is magnified 15 times by a voltage amplifier. Signals are then amplified and fed to a digital control system. The control algorithms are implemented using dSPACE DS1103 system with necessary Matlab/Simulink software installed in an industrial computer. The control algorithm is implemented using Simulink software and Real Time Workshop (RTW) is used to generate C code from the developed Simulink model. The C code is then converted to target specific code by real time interface (RTI) and target language compiler (TLC) supported by DS1103 controller board. Then we can design a vibration control experiment in real time by using ControlDestk software provided by dSPACE. The control objective is to minimize the output strain of two sensor outputs within the control bandwidth under the excitation of the disturbance force induced by external force.

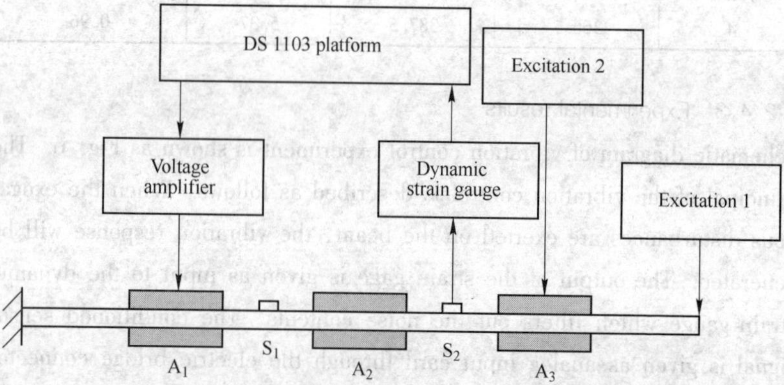

Fig. 6 Schematic diagram of vibration control experiment

The objective in the experimental studies is to control the first three vibration modes. In order to generate the disturbance force exerted on the beam, we use the actuator A_3 as excitation source. The actuators A_1, A_2 are applied to vibration control. The gain of the dynamic strain gauge is set to 1000, the bridge voltage 6V, cutoff frequency of filter is set to 100Hz. The sampling frequency of analog input and output port is set to 1000Hz. The covariance of measurement noise is estimated as 10^{-4}. The sampling period of the controller is 1ms. The real-time control time is set to 1s. The diagonal matrix δ in Eq. (22) is $\delta = \text{diag}\ (2\times10^3,\ 2\times10^3,\ 2\times10^3)$. The schematic diagram of the feedback control system is depicted in Fig. 6. In order to experimentally investigate the control performance, the two excitation sources are exerted to the cantilever beam, respectively. The first excitation source called excitation 1 is the free-vibration one by taping instantly the free end of the beam. The second excitation source named as excitation 2 is the forced one by applying a voltage to the piezoelectric ceramic patch A_3. The excitation voltage is generated by the signal generator which generates a white noise signal, amplified by voltage amplifier. And the covariance of the excitation signal is set to 120 V.

As we design the Kalman state estimator, the parameters are set as $Q_d = 100$, $R_e = 1\times10^{-3}$. In order to validate that the proposed estimator can effectively estimate the state values, we will carry out the following test. As the actuator A_3 is applied to a sinusoidal signal, we compared the measured values and estimated values of output strain from sensor S_1 and sensor S_2. The sinusoidal signal is generated by the signal generator, and is amplified by voltage amplifier, its amplitude 120 V, frequency 75 Hz, which the frequency is close to the second order natural one of the system to make the beam vibrate significantly. Fig. 7 and Fig. 8 denote the measured values

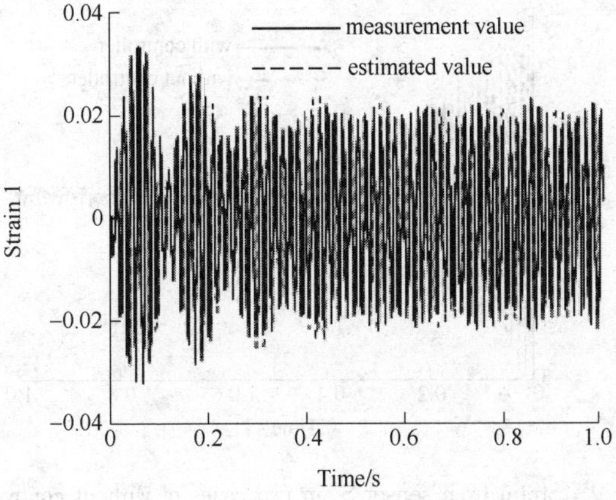

Fig. 7 Measurement and estimated values of output strain from sensor S_1

and estimated values of output strain from sensor S_1 and sensor S_2. From the two figures, we can observe the measurement values are coincide with estimated values of output strain to illustrate the effectiveness of the estimator.

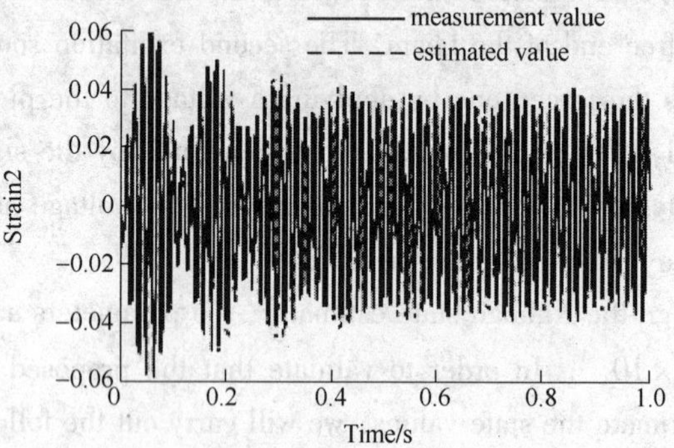

Fig. 8 Measurement and estimated values of output strain from sensor S_2

Fig. 9 and Fig. 10 show the strains of the flexible link from sensor S_1 and S_2 in two cases of without controller and with controller under the excitation 1, respectively. Fig. 11 and Fig. 12 show the strains of the flexible link

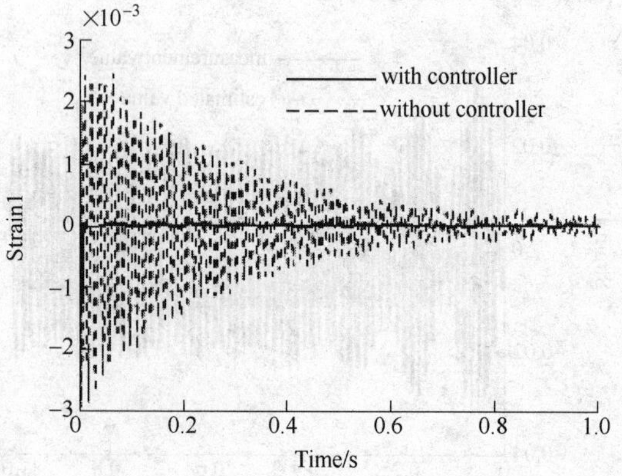

Fig. 9 Strain from sensor S_1 in two cases of without controller and with controller under the excitation 1

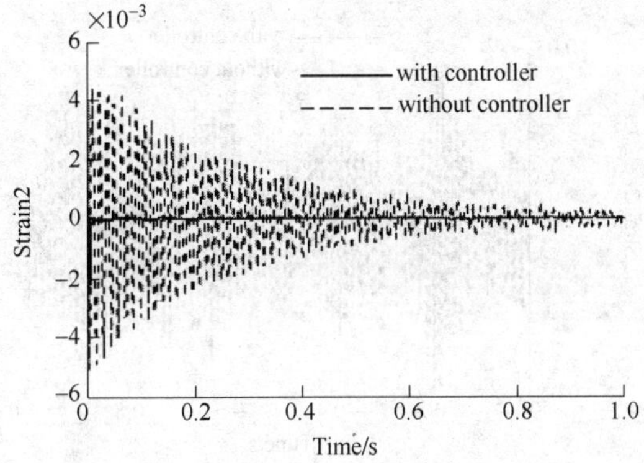

Fig. 10 Strain from sensor S_2 in two cases of without controller and with controller under the excitation 1

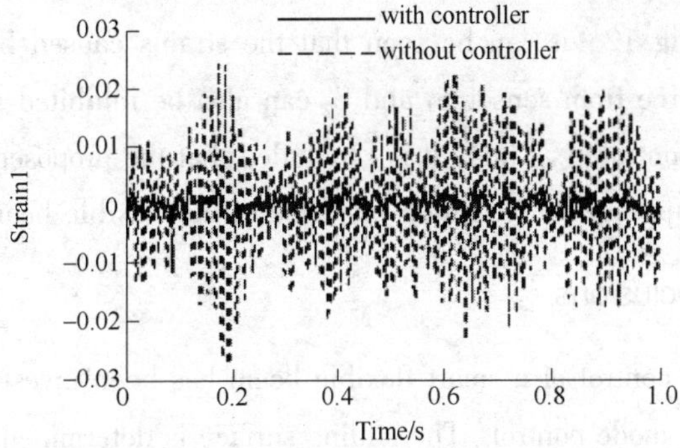

Fig. 11 Strain from sensor S_1 in two cases of without controller and with controller under the excitation 2

from sensor S_1 and S_2 in two cases of without controller and with controller under the excitation 2, respectively. In these figures, solid line represents the strains in case of with controller, and dotted line denotes the strains in case of without controller. By the contrast to the results of Fig. 9 and Fig. 10, it can be observed that the strains generated by the first excitation source from sensor S_1 and S_2 are remarkably reduced under the case where

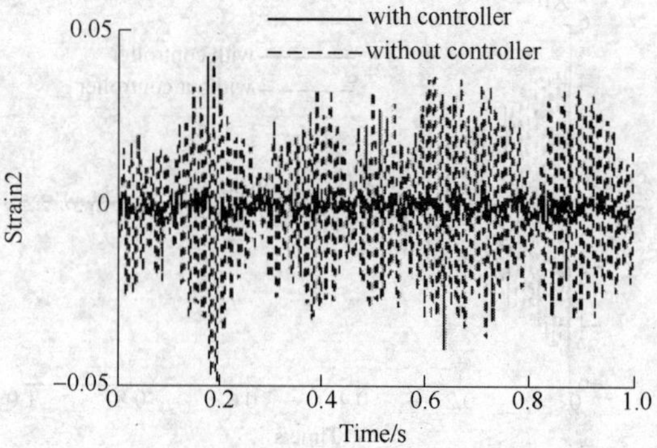

Fig. 12 Strain from sensor S_2 in two cases of without controller and with controller under the excitation 1

sliding mode controller is adopted. Comparing the experimental results of Fig. 11 and Fig. 12, it can be seen that the strains caused by the second excitation source from sensor S_1 and S_2 can also be inhibited greatly in the case of with controller. Thus, it is concluded that the proposed controller is effective to suppress the vibration response of the flexible beam.

4.2.5 Conclusions

The vibration control of a smart flexible beam has been investigated by applying sliding mode control. The sliding surface is determined by constructing a linear quadratic optimization problem, and Lyapunov direct method is used to design the sliding controller based on the Lyapunov asymptotic stability condition. The Kalman filter is applied considering uncertain disturbance and measurement noise. And the effectiveness of the proposed estimator was proved by comparing the measured and estimated values of the output strains. The experimental results have been showed that the proposed controller is valid to suppress the vibration response of the flexible beam. And it has been shown that the proposed controller guaranties robust performance.

References

[1] Choi S B. Active structural acoustic control of a smart plate featuring piezoelectric actuators [J]. Journal of Sound and Vibration, 2006, 294 (1~2): 421~429.

[2] Ma K. Adaptive nonlinear control of a clamped rectangular plate with PZT patches [J]. Journal of Sound and Vibration, 2003, 264 (4): 835~850.

[3] Ma K. Ghasemi-Nejhad M N. Frequency-weighted adaptive control for simultaneous precision positioning and vibration suppression of smart structures [J]. Smart Materials and Structures, 2004, 13 (5): 1143~1154.

[4] Abreu G L C M. Ribeiro J F, Steffen V. Experiments on optimal vibration control of a flexible beam containing piezoelectric sensors and actuators [J]. Shock and vibration, 2003, 10: 283~300.

[5] Li Y Y, Yam L H L. Robust vibration control uncertain systems using variable parameter feedback and model-based fuzzy strategies [J]. Computers and Structures, 2001, 79: 1109~1119.

[6] Mayhan P, Washington G. Fuzzy model reference learning control: a new control paradigm for smart structures [J]. Smart Mater. Struct., 1998, 7: 874~884.

[7] Zhang X M, Shao C J, Erdman A G. Active vibration controller design and comparison study of flexible linkage mechanism systems [J]. Mechanism & Machine Theory, 2002, 37: 985~997.

[8] Zhang X M, Shao C J, Li S, et al. Robust H_∞ vibration control for flexible linkage mechanism systems with piezoelectric sensors and actuators [J]. Journal of Sound and Vibration, 2001, 243 (1): 145~155.

[9] Kawabe H, Tsukiyama N, Yoshida K. Active vibration damping based on neural network theory [J]. Materials Science and Engineering A, 2006, 442: 547~550.

[10] Liao C Y, Sung C K. An elastodynamic analysis and control of flexible linkages using piezoceramic sensors and actuators [J]. Transactions of the ASME, 1993, 115: 658~665.

[11] Iorga L, Baruh H, Ursu I. A Review of H_∞ robust control of piezoelectric smart structures [J]. Appl. Mech. Rev., 2008, 61 (4): 1~16.

4.3 柔性并联机器人动力学建模

本文针对一般柔性并联机器人动力学模型，提出了一种精确而简单的动力学建模方法。根据并联机器人结构特点，将其划分为若干刚性子结构和弹性子结构，形成一个刚柔结合的系统。静平台和动平台相对其他构件变形较小，将其作为刚性子结构，各个支链作为弹性子结构。分别建立各子结构的动力学方程，弹性子结构采用有限元方法和模态综合法建立其动力学方程；考虑各个柔性支链弹性变形对刚性子结构的影响，建立刚性子结构动力学方程；推导出相邻的刚性子结构和弹性子结构之间的几何约束关系。通过相邻子结构的协调矩阵，将各个子结构的方程进行装配形成系统的弹性动力学方程。通过一种高速并联机械手的动力学特性比较分析，表明该方法的正确性和可行性。由于引入刚性子结构和采用了模态综合法，减少了系统运动关联性，从而简化了计算模型。

4.3.1 引言

并联机器人具有高速度 高精度 高承载能力等特点，目前，对并联机器人的研究主要基于运动学和刚体动力学分析，工程实践表明，这种分析方法对于中、低速系统通常是适宜的。现代机械向高速、精密、轻型等方向发展，必须考虑并联机器人中构件的变形。因为构件的弹性变形影响原设计的精度并会导致整个机器人的冲击、噪声和疲劳，所以柔性并联机器人的研究具有重要的理论和实际意义[1,2]。

柔性并联机器人是一个多闭环、多柔体的非线性动力学系统，其动力学方程的建立和动力学性能的分析较困难[3~8]。柔性并联机器人动力学建模方面的研究还很不足，为了更有效地进行弹性动力学分析，设计出动力学性能最优的并联机器人和主动振动控制器，必须建立精确而简单的动力学模型。

本文以一般的柔性并联机器人为研究对象，根据并联机器人的结构特点，将其划分成一个刚体和柔性体结合的系统，分别建立它们的动力学方程，推导它们之间的几何约束关系，推导系统的弹性动力学方程。

4.3.2 系统动力学方程

柔性并联机构是一个多闭环、多柔体的非线性动力学系统，其动力学建模比刚性并联机器人和柔性串联机器人复杂。在建模过程中，应考虑以下因素：并联机构的结构特点是由若干独立运动支链、静平台和动平台组成。在柔性并联机构中，为了简化模型视某些构件为刚体，所以它是由刚体和弹性体组成的多体系统。将结构尺寸大、刚度高的构件或部件作为刚性子结构；将有较大弹性变形的杆件或杆件组作为弹性子结构。根据轻型并联机构的结构和运动特点，一般其静平台和动平台相对于其他运动构件尺寸更大、刚度更高，而各个独立的运动支链一般是由细长杆组成，所以可以选择静平台和动平台作为刚性结构，各个运动支链为弹性子结构。

4.3.2.1 单元动力学方程

机器人的弹性构件用梁单元模拟。考虑刚体运动与弹性变形运动的耦合和几何刚化的影响，采用有限元法和浮点坐标法建立单元运动微分方程[9]

$$\overline{\boldsymbol{m}}\ddot{u}+2\dot{\theta}b\dot{u}+(\overline{\boldsymbol{k}}_1+\overline{\boldsymbol{k}}_2+\ddot{\theta}b-\dot{\theta}^2\overline{\boldsymbol{m}})u=\overline{\boldsymbol{p}} \tag{1}$$

式中，$\overline{\boldsymbol{m}}$为单元质量矩阵；$\overline{\boldsymbol{k}}_1$为单元刚度矩阵；$\overline{\boldsymbol{p}}$为广义力；$2\dot{\theta}b$ 为陀螺阻尼；$\overline{\boldsymbol{k}}_2$为几何刚化矩阵；$\ddot{\theta}b-\dot{\theta}^2\overline{\boldsymbol{m}}$为离心刚度。

4.3.2.2 弹性支链的动力学方程

将支链划分为有限个单元，根据有限元叠加原理，将单元动力学方程组合成子结构的动力学方程

$$\boldsymbol{M}_i\ddot{\boldsymbol{U}}_i+\boldsymbol{C}_i\dot{\boldsymbol{U}}_i+\boldsymbol{D}_i\boldsymbol{U}_i=\boldsymbol{P}_i \tag{2}$$

式中，$\boldsymbol{M}_i$ 为支链等效质量矩阵；$\boldsymbol{C}_i$ 为支链等效阻尼矩阵；$\boldsymbol{D}_i$ 为支链等效刚度矩阵；$\boldsymbol{P}_i$ 为广义力；$\boldsymbol{U}_i$ 为子结构的广义坐标。

由于广义坐标数目较大，采用约束模态综合法缩减坐标。设 $\boldsymbol{U}_{ri}$ 为缩减后的坐标，它由支链的边界坐标与截断振型坐标组成，$\boldsymbol{A}$ 为模态变换矩阵[9]，则有：

$$\boldsymbol{U}_i=\boldsymbol{A}\boldsymbol{U}_{ri} \tag{3}$$

将式（3）代入式（2），并前乘 $\boldsymbol{A}^{\mathrm{T}}$ 得到坐标缩减后的子结构动力学方程为：

$$\boldsymbol{M}_{ri}\ddot{\boldsymbol{U}}_{ri}+\boldsymbol{C}_{ri}\dot{\boldsymbol{U}}_{ri}+\boldsymbol{D}_{ri}\boldsymbol{U}_{ri}=\boldsymbol{P}_{ri} \tag{4}$$

式中，$\boldsymbol{M}_{ri}=\boldsymbol{A}^{\mathrm{T}}\boldsymbol{M}_i\boldsymbol{A}$，$\boldsymbol{C}_{ri}=\boldsymbol{A}^{\mathrm{T}}\boldsymbol{C}_i\boldsymbol{A}$，$\boldsymbol{D}_{ri}=\boldsymbol{A}^{\mathrm{T}}\boldsymbol{D}_i\boldsymbol{A}$，$\boldsymbol{P}_{ri}=\boldsymbol{A}^{\mathrm{T}}\boldsymbol{P}_i$

4.3.2.3 几何约束关系

由于弹性子结构的弹性变形会使刚性子结构产生位置误差，为了将各子结构运动方程装配成系统的弹性动力学方程和减少系统自由度

数，必须建立它们之间的协调矩阵，建立该协调矩阵的原则是已知弹性子结构与刚性子结构联接点的变形量来求解刚性子结构质心的位置误差，而且协调矩阵是关于它们独立的，这样便于得出它们之间的速度和加速度关系。如图1所示，1为刚体，2为某一弹性子结构与刚性子结构相连的弹性构件，二者在 A 点以运动副相连。刚体的质心为 C。在建立各弹性子结构方程时，弹性构件2在 A 点处有6个广义坐标 $\boldsymbol{U}_A=[\Delta X_A \quad \Delta Y_A \quad \Delta Z_A \quad \Delta\theta_{Ax} \quad \Delta\theta_{Ay} \quad \Delta\theta_{Az}]^{\mathrm{T}}$，前3个为位移，后3个为转角。刚体的广义坐标是由柔性支链的弹性变形在质心 C 产生的微位移和微转角，表示为 $\boldsymbol{U}_C=[\Delta X_C \quad \Delta Y_C \quad \Delta Z_C \quad \Delta\theta_{Cx} \quad \Delta\theta_{Cy} \quad \Delta\theta_{Cz}]^{\mathrm{T}}$ 质心 C 由理论位置运动到 C'，A 点运动到 A' 点。下面推导它们之间的几何关系

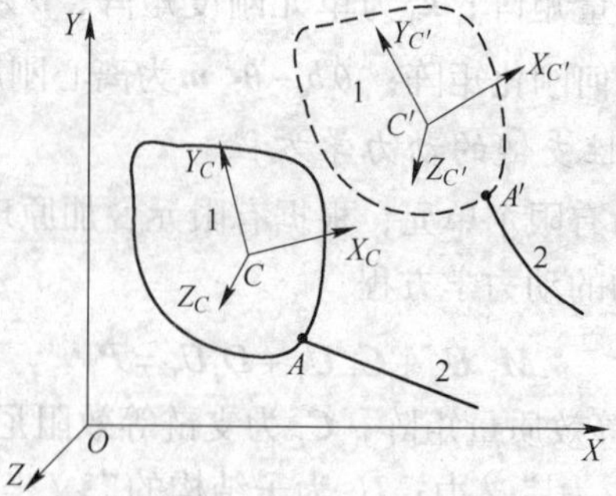

图1 几何约束关系

为了描述它们的关系，定义3个坐标系，$O-XYZ$ 为总体坐标系，$C-X_CY_CZ_C$ 为原点在 C 的随刚体一起运动的局部坐标系，$C'-X_{C'}Y_{C'}Z_{C'}$ 为原点在 C' 的局部坐标系。设 $\boldsymbol{R}$ 为 $C-X_CY_CZ_C$ 相对于总体坐标系的广义变换矩阵，$\boldsymbol{R}$ 为 $C'-X_{C'}Y_{C'}Z_{C'}$ 相对于 $C-X_CY_CZ_C$ 的广义变换矩阵，$\boldsymbol{R}'$ 为 $C'-X_{C'}Y_{C'}Z_{C'}$ 相对于 $O-XYZ$ 的广义变换矩阵，它们分别为

$$\boldsymbol{R}=\begin{bmatrix} \cos\theta_{C_x} & -\sin\theta_{C_z} & \sin\theta_{C_y} & X_C \\ \sin\theta_{C_z} & \cos\theta_{C_y} & -\sin\theta_{C_x} & Y_C \\ -\sin\theta_{C_y} & \sin\theta_{C_x} & \cos\theta_{C_z} & Z_C \\ 0 & 0 & 0 & 1 \end{bmatrix} \tag{5}$$

式中，θ_{C_i}为坐标系与总坐标系坐标轴的夹角；（X_C，Y_C，Z_C）为 C 点在总坐标系下的坐标。

由于弹性变形小，质心 C 产生的微位移和微转角较小，则可假设：

$$\cos\Delta\theta_{C_i} \approx 1, \sin\Delta\theta_{C_i} \approx \Delta\theta_{C_i} \tag{6}$$

则 $C' - X_{C'}Y_{C'}Z_{C'}$ 相对于 $C - X_C Y_C Z_C$ 的广义变换矩阵为：

$$\Delta \boldsymbol{R} = \begin{bmatrix} 1 & -\Delta\theta_{C_z} & \Delta\theta_{C_y} & \Delta X_C \\ \Delta\theta_{C_z} & 1 & -\Delta\theta_{C_x} & \Delta Y_C \\ -\Delta\theta_{C_y} & \Delta\theta_{C_x} & 1 & \Delta Z_C \\ 0 & 0 & 0 & 1 \end{bmatrix} \tag{7}$$

$C' - X_{C'}Y_{C'}Z_{C'}$ 相对于 $O - XYZ$ 的广义变换矩阵为：

$$\boldsymbol{R}' = \Delta \boldsymbol{R}\boldsymbol{R} \tag{8}$$

设 A 点和 A' 点的坐标分别用 $[X_A \quad Y_A \quad Z_A]^{\mathrm{T}}$ 和 $[X'_A \quad Y'_A \quad Z'_A]^{\mathrm{T}}$ 来表示。由坐标变换可得：

$$\begin{bmatrix} X'_A \\ Y'_A \\ Z'_A \\ 1 \end{bmatrix}_O = \boldsymbol{R}' \begin{bmatrix} X'_A \\ Y'_A \\ Z'_A \\ 1 \end{bmatrix}_{C'} \tag{9}$$

式中，下标 O 和 C' 分别表示在 $O - XYZ$ 和 $C' - X_{C'}Y_{C'}Z_{C'}$ 下的坐标。由

$$\begin{bmatrix} X'_A \\ Y'_A \\ Z'_A \\ 1 \end{bmatrix}_{C'} = \boldsymbol{R}' \begin{bmatrix} X'_A \\ Y'_A \\ Z'_A \\ 1 \end{bmatrix}_C \tag{10}$$

则有：

$$\begin{bmatrix} X'_A \\ Y'_A \\ Z'_A \\ 1 \end{bmatrix}_O = \Delta \boldsymbol{R}\boldsymbol{R} \begin{bmatrix} X_A \\ Y_A \\ Z_A \\ 1 \end{bmatrix}_C = \Delta \boldsymbol{R} \begin{bmatrix} X_A \\ Y_A \\ Z_A \\ 1 \end{bmatrix}_O \tag{11}$$

$$\begin{bmatrix} \Delta X_A \\ \Delta Y_A \\ \Delta Z_A \\ 1 \end{bmatrix}_O = \begin{bmatrix} X'_A \\ Y'_A \\ Z'_A \\ 1 \end{bmatrix}_O - \begin{bmatrix} X_A \\ Y_A \\ Z_A \\ 1 \end{bmatrix}_O = (\Delta \boldsymbol{R} - I)\begin{bmatrix} X_A \\ Y_A \\ Z_A \\ 1 \end{bmatrix}_O \tag{12}$$

将式（8）代入式（12）并整理得到：

$$\begin{bmatrix} \Delta X_A \\ \Delta Y_A \\ \Delta Z_A \end{bmatrix} = [\boldsymbol{I} \quad \boldsymbol{T}_A]\begin{bmatrix} \Delta X_C \\ \Delta Y_C \\ \Delta Z_C \\ \Delta\theta_{C_x} \\ \Delta\theta_{C_y} \\ \Delta\theta_{C_z} \end{bmatrix} \tag{13}$$

式中，$\boldsymbol{T}_A = \begin{bmatrix} 0 & Z_A & -Y_A \\ -Z_A & 0 & X_A \\ Y_A & -X_A & 0 \end{bmatrix}$，$\boldsymbol{I}$ 为单位矩阵。

A 点的转动角位移等于 C 点的角位移，即

$$[\Delta\theta_{A_x} \quad \Delta\theta_{A_y} \quad \Delta\theta_{A_z}]^{\mathrm{T}} = I[\Delta\theta_{C_x} \quad \Delta\theta_{C_y} \quad \Delta\theta_{C_z}]^{\mathrm{T}} \tag{14}$$

综合上两式可得到刚性子结构和弹性子结构的位移协调矩阵：

$$\boldsymbol{U}_A = \boldsymbol{B}\boldsymbol{U}_C \tag{15}$$

式中，$\boldsymbol{B} = \begin{bmatrix} \boldsymbol{I} & \boldsymbol{T}_A \\ \boldsymbol{0} & \boldsymbol{I} \end{bmatrix}$，$\boldsymbol{I}$ 和 $\boldsymbol{0}$ 分别为 3×3 的单位矩阵和零矩阵，$\boldsymbol{B}$ 为位移协调矩阵。$\boldsymbol{T}_A$ 只和 A 点的坐标有关，则可知刚性子结构和弹性子结构之间的位移协调矩阵 $\boldsymbol{B}$ 是关于 $\boldsymbol{U}_A$ 和 $\boldsymbol{U}_C$ 独立的，也即是矩阵 $\boldsymbol{B}$ 中不含变量 $\boldsymbol{U}_A$ 和 $\boldsymbol{U}_C$。因此，速度和加速度的关系为：

$$\dot{\boldsymbol{U}}_A = \boldsymbol{B}\dot{\boldsymbol{U}}_C + \dot{\boldsymbol{B}}\boldsymbol{U}_C \tag{16}$$

$$\ddot{\boldsymbol{U}}_A = \boldsymbol{B}\ddot{\boldsymbol{U}}_C + \dot{\boldsymbol{B}}\dot{\boldsymbol{U}}_C + \dot{\boldsymbol{B}}\dot{\boldsymbol{U}}_C + \ddot{\boldsymbol{B}}\boldsymbol{U}_C \tag{17}$$

4.3.2.4 刚性子结构动力学方程

设 C'点和 C 点在总坐标系下的位移和转角分别为 $\boldsymbol{P}_{C'}$、$\boldsymbol{P}_C$，表示为：

$$\begin{cases} \boldsymbol{P}_{C'} = [X_{C'} \quad Y_{C'} \quad Z_{C'} \quad \theta_{C'_x} \quad \theta_{C'_y} \quad \theta_{C'_z}]^{\mathrm{T}} \\ \boldsymbol{P}_C = [X_C \quad Y_C \quad Z_C \quad \theta_{C_x} \quad \theta_{C_y} \quad \theta_{C_z}]^{\mathrm{T}} \end{cases} \tag{18}$$

同理，有如下关系：

$$\boldsymbol{P}_{C'} = \boldsymbol{P}_C + \boldsymbol{B}_C \boldsymbol{U}_C \tag{19}$$

式中，$\boldsymbol{B}_C = \begin{bmatrix} \boldsymbol{I} & \boldsymbol{T}_C \\ \boldsymbol{0} & \boldsymbol{I} \end{bmatrix}$，$\boldsymbol{T}_C = \begin{bmatrix} 0 & Z_C & -Y_C \\ -Z_C & 0 & X_C \\ Y_C & -X_C & 0 \end{bmatrix}$。

对上式求导，可得 C' 点的速度：

$$\dot{\boldsymbol{P}}_{C'} = \dot{\boldsymbol{P}}_C + \dot{\boldsymbol{B}}_C \boldsymbol{U}_C + \boldsymbol{B}_C \dot{\boldsymbol{U}}_C \tag{20}$$

则刚体的动能可表示为：

$$T = \frac{1}{2} \dot{\boldsymbol{P}}_{C'}^{\mathrm{T}} \boldsymbol{M}^* \dot{\boldsymbol{P}}_{C'} \tag{21}$$

式中，$\boldsymbol{M}^* = \begin{bmatrix} m_C & 0 & 0 & 0 & 0 & 0 \\ 0 & m_C & 0 & 0 & 0 & 0 \\ 0 & 0 & m_C & 0 & 0 & 0 \\ 0 & 0 & 0 & I_x & 0 & 0 \\ 0 & 0 & 0 & 0 & I_y & 0 \\ 0 & 0 & 0 & 0 & 0 & I_z \end{bmatrix}$，$m_C$ 为刚性子结构质量；I_x 为绕 x 轴转动惯量；I_y 为绕 y 轴转动惯量；I_z 为绕 z 轴转动惯量。

刚体重力势能为：

$$\boldsymbol{U} = [0 \quad 0 \quad 1 \quad 0 \quad 0 \quad 0] \boldsymbol{P}_{C'} \tag{22}$$

设刚性子结构的广义坐标为 $\boldsymbol{U}_C$，经整理后可得刚性子结构的动力学方程为：

$$\boldsymbol{M}_R \ddot{\boldsymbol{U}}_C + \boldsymbol{C}_R \dot{\boldsymbol{U}}_C + \boldsymbol{D}_R \boldsymbol{U}_C = \boldsymbol{F}_R - \boldsymbol{Q}_R \tag{23}$$

式中，$\boldsymbol{M}_R = \boldsymbol{B}_C^{\mathrm{T}} \boldsymbol{M}^* \boldsymbol{B}_C$，$\boldsymbol{C}_R = 2\boldsymbol{B}_C^{\mathrm{T}} \boldsymbol{M}^* \dot{\boldsymbol{B}}_C$，$\boldsymbol{D}_R = \boldsymbol{B}_C^{\mathrm{T}} \boldsymbol{M}^* \ddot{B}_C$，$\boldsymbol{Q}_R = \boldsymbol{B}_C^{\mathrm{T}} \boldsymbol{M}^* \ddot{\boldsymbol{P}}_C + mg\boldsymbol{B}_C^{\mathrm{T}}$。$\boldsymbol{F}_R$ 为外力矢量等效到质心处的力矢量。

4.3.2.5 系统弹性动力学方程

设有 i 个弹性支链，根据支链的动力学方程可知，各个支链的广义坐标为 $\boldsymbol{U}_{ri}$，包括 $\boldsymbol{U}_A$ 等6 个广义坐标，由于存在约束关系，则选取系统广义坐标为：

$$\boldsymbol{U} = [U_1 \quad \cdots \quad U_{6-i} \quad \boldsymbol{U}_C]^{\mathrm{T}} \tag{24}$$

U_{6-i}表示第 i 个支链所有广义坐标，广义坐标包含动平台的位置误差，便于求解。各个子结构广义坐标与系统广义坐标的关系为：

$$\boldsymbol{U}_{ri}=\boldsymbol{B}_i\boldsymbol{U},\ \boldsymbol{U}_C=\boldsymbol{B}_{i+1}\boldsymbol{U}$$

将上式代入各子结构动力学方程并前乘 $\boldsymbol{B}_j$，所有方程叠加可得系统动力学方程为：

$$\boldsymbol{M}\ddot{\boldsymbol{U}}+\boldsymbol{C}\dot{\boldsymbol{U}}+\boldsymbol{D}\boldsymbol{U}=\boldsymbol{P} \tag{25}$$

其中，

$$\boldsymbol{M}=\sum_{j=1}^{i+1}\boldsymbol{B}_j^{\mathrm{T}}\boldsymbol{M}_j\boldsymbol{B}_j$$

$$\boldsymbol{P}=\sum_{j=1}^{i+1}\boldsymbol{B}_j^{\mathrm{T}}\boldsymbol{P}_j$$

$$\boldsymbol{C}=\sum_{j=1}^{i+1}\boldsymbol{B}_j^{\mathrm{T}}\boldsymbol{C}_j\boldsymbol{B}_j+2\boldsymbol{B}_j^{\mathrm{T}}\boldsymbol{M}_j\dot{\boldsymbol{B}}_j$$

$$\boldsymbol{D}=\sum_{j=1}^{i+1}\boldsymbol{B}_j^{\mathrm{T}}\boldsymbol{D}_j\boldsymbol{B}_j+\boldsymbol{B}_j^{\mathrm{T}}\boldsymbol{M}_j\ddot{\boldsymbol{B}}_j+\dot{\boldsymbol{B}}_j^{\mathrm{T}}\boldsymbol{C}_j\dot{\boldsymbol{B}}_j$$

式中，$\boldsymbol{M}$ 为系统等效质量矩阵；$\boldsymbol{C}$ 系统等效阻尼矩阵；$\boldsymbol{D}$ 为系统刚度矩阵；$\boldsymbol{P}$ 为系统广义力。

4.3.3 算例

为了验证该建模方法的有效性，以实验室高速机械手为算例[10]。该机械手结构如图 2 所示，由 2 条主动支链、2 条从动支链、1 个动

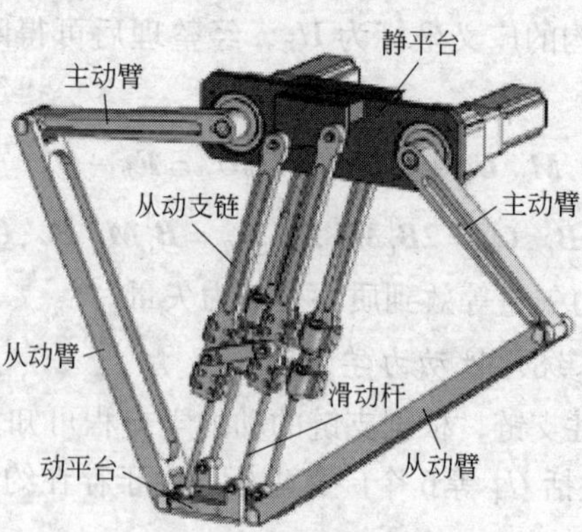

图 2 并联机械手

平台和1个静平台组成。每条主动支链含有1条主动臂和1条从动臂。机械手各构件的材料是铝合金，弹性模量和密度分别为6.94×10^4MPa、2.7×10^3kg/m^3。机械手的动平台质量为0.30kg，机械手的主动臂、从动臂和从动支链横截面为“工”字形，滑动杆长度为365mm、直径为8mm的园杆。机构的结构参数如表1所示。

表1 机械手结构参数

构 件	长度/mm	截面参数/mm			
		高度	翼缘厚	翼缘宽	腹板厚
主动臂	280	30	8	30	5
从动臂	505	60	4	27	4
从动支链	360	23	4.5	15	5

根据该并联机械手的结构特点，将其划分为由主动臂和从动臂组成的2个弹性支链，由从动支链与滑动杆组成的4个弹性支链，总共6个弹性支链，动平台作为刚性子结构。为保证计算精度，每个弹性支链分为10个单元，整个机构共60个单元，61个节点，由于该机械手为平面机构，每个节点有3个广义坐标。如图3所示，采用模态综合法，每一个弹性支链取其边界处的坐标为4个，取各自前3阶截断振型坐标，则每个弹性支链有7个坐标，6个弹性支链共42个坐标。由于用动平台质心的微位移和微转角 U_C 作为广义坐标，它与弹性支链点 A_i 处的广义坐标存在文中所表述的几何约束关系，则整个系统的广义坐标只有27个，这样减少了系统坐标数量，简化了数学

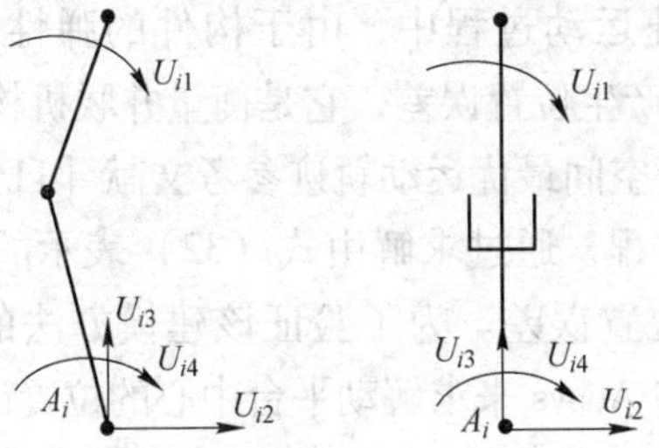

图3 弹性支链广义坐标

模型。按上述方法分别建立各个支链的动力学方程，并得到机械手系统动力学方程。

为了验证该方法的有效性和正确性，利用该方法得到的系统动力学方程进行机械手的动力学特性比较分析。首先，分析比较其固有频率，第1种方法是使用式（32）的质量矩阵和刚度矩阵计算系统固有频率，第2种方法是使用 Ansys Workbench 对机械手进行模态分析得到其固有频率，机械手的网格划分如图4所示，为了与第1种方法比较，在网格划分时，静平台和动平台均设为刚体，其他构件的联结关系设为相应的运动副。使用这两种方法分别计算了机械手在平衡位置时的前3阶固有频率，比较结果如表2所示。从表2可以看出，前3阶的固有频率的相对误差分别为4.3%、4.8%、6.4%，可见两种方法计算的结果比较接近，这说明该方法是可行的。

图4 机械手的网格划分

机械手在高速运动过程中，由于构件的弹性变形机械手动平台（末端执行器）会产生位置误差，它是衡量并联机构精度的一个重要指标。机械手的关节空间最优运动轨迹参考文献［11］。根据前面系统动力学方程的建立过程，通过求解由式（32）表示的系统微分方程可得到动平台质心的位置误差。为了验证该建模方法的有效性和正确性，分别运用本方法和 Ansys 来求解动平台中心的位置误差。图5和图6分别表示使用这两种方法动平台中心在 XY 方向的位置误差。由图5和图6可知，这两种方法得到的结果较接近，证明该方法正确。

表2 固有频率比较结果

固有频率	本方法计算结果/Hz	Ansys 计算结果/Hz	相对误差/%
第1阶	74.362	75.951	4.3
第2阶	98.874	94.256	4.8
第3阶	173.846	166.650	6.4

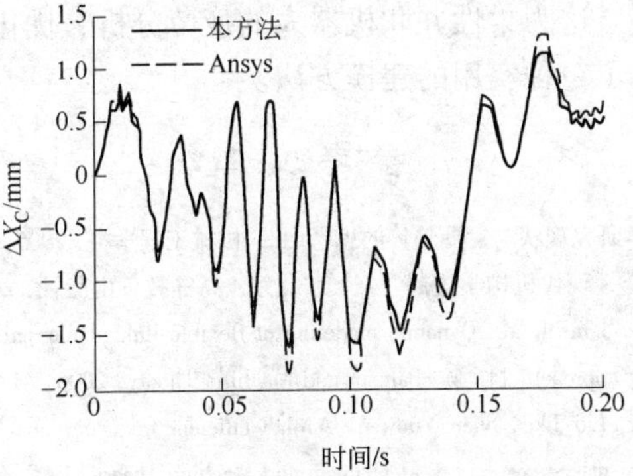

图5 动平台质心的 X 方向位置误差

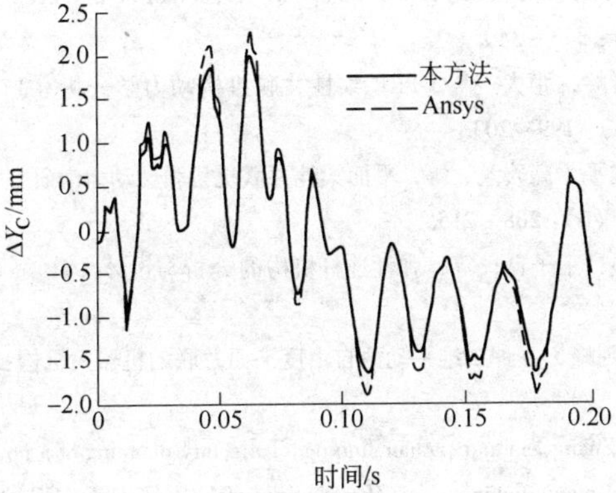

图6 动平台质心的 Y 方向位置误差

4.3.4 结论

（1）根据并联机构是由若干独立运动支链、静平台和动平台组成的特点，引入刚性子结构，推导出相邻的刚性子结构和弹性子结构之间的几何约束关系。使用动平台质心的微位移和微转角作为系统广义坐标并采用了模态综合法，减少了系统运动参数，简化了模型。

（2）通过一种高速并联机械手的动力学特性比较分析，表明该方法是可行的，而且建模方便、直观，便于求解并联机构的末端执行器的位置误差。这为柔性并联机器人的疲劳分析、优化设计、振动控制等研究提供了一种实用的建模方法。

参考文献

[1] 高峰. 机构学研究现状与发展趋势的思考[J]. 机械工程学报, 2005, 41(8): 3 ~ 17.

[2] 邹慧君, 高峰. 现代机构学进展 [M]. 北京: 高等教育出版社, 2007.

[3] Wang Xiaoyun, Jame K M. Dynamic modeling of flexible-link planar parallel platform using a sub-structuring approach[J]. Mechanism and Machine Theory, 2006, 41(6): 671 ~ 678.

[4] Zhang Xianmin, Liu Jike, Shen Yunwen. A high efficient frequency analysis method for closed flexible mechanism systems [J]. Mechanism and Machine Theory, 1998, 33 (8): 1117 ~ 1125.

[5] Zhang Xianmin, Liu Hongzhao, Shen Yunwen. Finite dynamic element analysis for high-speed flexible linkage mechanisms [J]. Computers & Structures, 1996, 60 (5): 787 ~ 796.

[6] Zhang Xuping, James K M. Dynamic modeling and experimental validation of 3-PRR parallel manipulator with flexible intermediate links [J]. Journal of Intelligent & Robotic Systems, 2007, 50 (4): 323 ~ 340.

[7] 贾晓辉, 田延岭, 张大卫. 3-PRR 柔性并联机构动力学分析 [J]. 农业机械学报, 2010, 41 (10): 199 ~ 203.

[8] 鲁开讲, 师俊平, 高秀兰, 等. 平面柔性并联机构弹性动力学研究 [J]. 农业机械学报, 2010, 41 (6): 208 ~ 215.

[9] 张策, 黄永强, 王子良, 等. 弹性连杆机构的分析与设计 [M]. 北京: 机械工业出版社, 1997.

[10] 张宪民, 袁剑峰. 一种二维平动两自由度平面并联的机器人机构: 中国, CN 1903521 [P], 2008.

[11] Hu Junfeng, Zhang Xianmin, Zhan Jinqing. Trajectory planning of a novel 2-DoF high-speed planar parallel manipulator [C] //Proceedings of 1st International Conference on Intelligent Robotics and Applications, ICIRA 2008, Wuhan, China, 2008: 199 ~ 207.

冶金工业出版社部分图书推荐

书　名	作　者	定价(元)
数控机床操作与维修基础(本科教材)	宋晓梅	29.00
自动控制系统(第2版)(本科教材)	刘建昌	15.00
少自由度并联支撑机构动基座自动调平系统	朱大昌	19.00
可编程序控制器及常用控制电器(第2版)(本科教材)	何友华	30.00
机电一体化技术基础与产品设计(第2版)(本科教材)	刘　杰	46.00
自动控制原理(第4版)(本科教材)	王建辉	32.00
自动控制原理习题详解(本科教材)	王建辉	18.00
机械设计基础(本科教材)	王健民	40.00
机械优化设计方法(第3版)(本科教材)	陈立周	29.00
现代机械设计方法(本科教材)	臧　勇	22.00
机械可靠性设计(本科教材)	孟宪铎	25.00
液压传动与气压传动(本科教材)	朱新才	39.00
计算机控制系统(本科教材)	张国范	29.00
冶金设备及自动化(本科教材)	王立萍	29.00
机械制造工艺及专用夹具设计指导(第2版)(本科教材)	孙丽媛	20.00
机械电子工程实验教程(本科教材)	宋伟刚	29.00
复杂系统的模糊变结构控制及其应用	米　阳	20.00
液压可靠性与故障诊断(第2版)	湛从昌	49.00
液压气动技术与实践(高职教材)	胡运林	39.00
机械制图(高职教材)	阎　霞	30.00
机械制图习题集(高职教材)	阎　霞	29.00
机械制造装备设计	王启义	35.00
带式输送机实用技术	金丰民	59.00
冶金通用机械与冶炼设备	王庆春	45.00
电气设备故障检测与维护	王国贞	28.00
机器人技术基础	柳洪义	23.00